Stories
FROM THE FIELD

YDF FOOTPRINT IN AFRICA

Youth Development through Football (YDF)
& Professor Cora Burnett, University of Johannesburg

Published in 2012 by
GIZ/ Youth Development through Football (YDF)
in collaboration with the University of Johannesburg

ISBN: 978-1-920143-77-0

Design and origination by SA Publishing Services
Cover design by Kerrin Cocks
Printed and bound by Pinetown Printers, Durban

The Youth Development through Football (YDF) project is implemented by the Deutsche Gesellschaft für Internationale Zusammenarbeit (GIZ) GmbH on behalf of the German Federal Government in cooperation with the Department of Sport and Recreation South Africa (SRSA). YDF is funded by the German Federal Ministry for Economic Cooperation and Development (BMZ) and co-funded by the European Union (EU).

Implemented by GIZ

On behalf of the German Government

Co-funded by the European Union

In partnership with SRSA

Contents

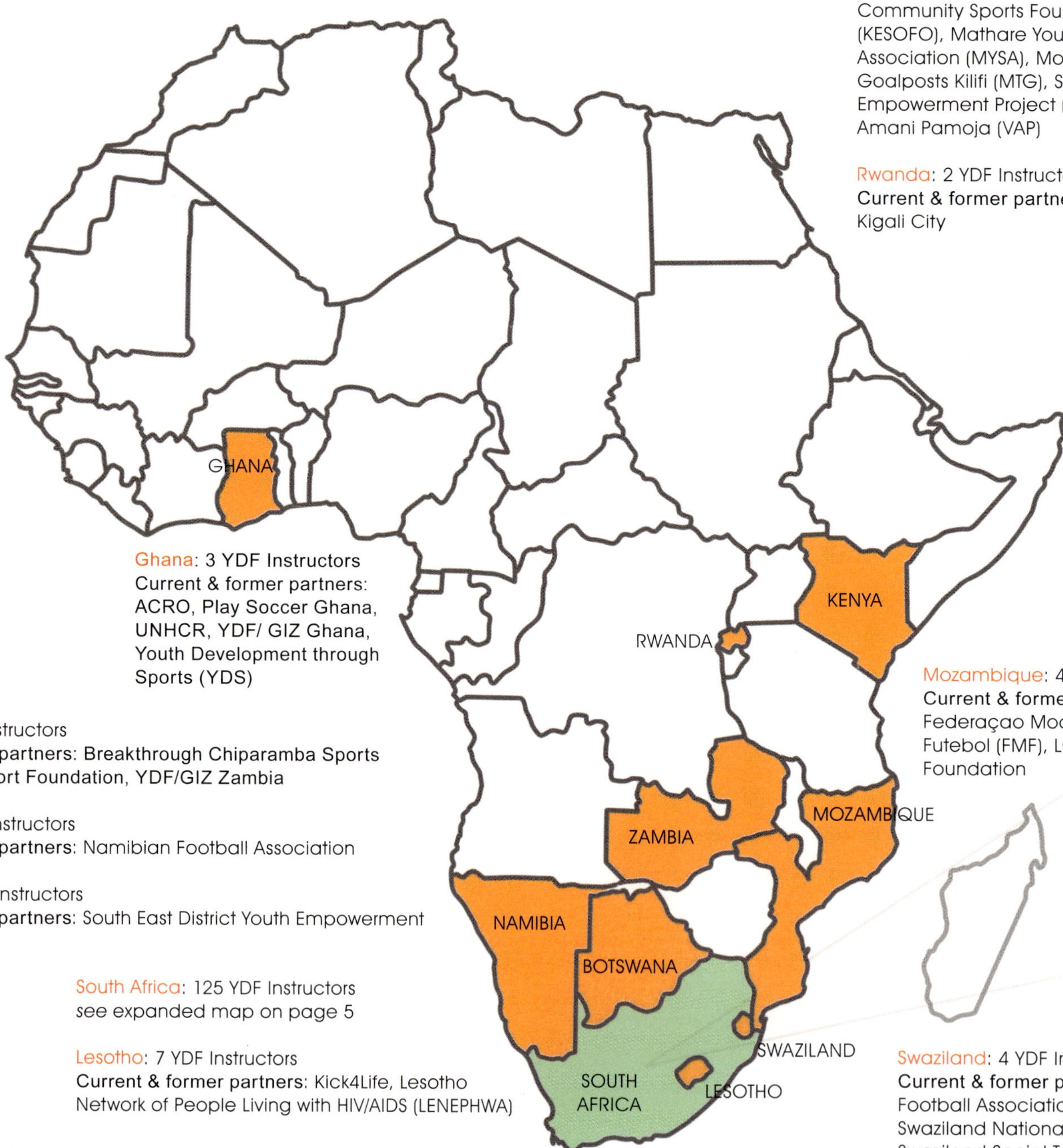

Kenya: 2 YDF Instructors
Current & former partners: Kenya Community Sports Foundation (KESOFO), Mathare Youth Sport Association (MYSA), Moving the Goalposts Kilifi (MTG), Society Empowerment Project (SEP), Vijana Amani Pamoja (VAP)

Rwanda: 2 YDF Instructors
Current & former partners: Esperance Kigali City

Ghana: 3 YDF Instructors
Current & former partners: ACRO, Play Soccer Ghana, UNHCR, YDF/ GIZ Ghana, Youth Development through Sports (YDS)

Mozambique: 4 YDF Instructors
Current & former partners: Federaçao Mocambicana de Futebol (FMF), Lurdes Mutola Foundation

Zambia: 3 YDF Instructors
Current & former partners: Breakthrough Chiparamba Sports Academy, EduSport Foundation, YDF/GIZ Zambia

Namibia: 5 YDF Instructors
Current & former partners: Namibian Football Association

Botswana: 3 YDF Instructors
Current & former partners: South East District Youth Empowerment League (SEDYEL)

South Africa: 125 YDF Instructors
see expanded map on page 5

Lesotho: 7 YDF Instructors
Current & former partners: Kick4Life, Lesotho Network of People Living with HIV/AIDS (LENEPHWA)

Swaziland: 4 YDF Instructors
Current & former partners: National Football Association of Swaziland, Swaziland National Sports Council, Swaziland Social Transformation Network (SSTN)

North West: 8 YDF Instructors
Current & former partners: Early Birds Football Club, Leseding Community Development Project

Gauteng: 37 YDF Instructors
Current & former partners: Altus Sport, Cometsa, Conquest for Life, Community Works Programme (CWP) Erasmus, CWP Merafong, Dlala Ntombazane, Dona's Mates, Greenfeet, Karos & Kambro, Play Soccer South Africa, SA Cares for Life/ Lesedi La Batho, SAFA Metsweding, SAFA Sedibeng, Southern Africa Social Development Agency, Sport for Social Change Network (SSCN) Southern Africa, Youth Sport for Social Development (YSSD)

Free State: 9 YDF Instructors
Current & former partners: Mangaung University of Free State Community Partnership Programme (MUCPP), SAFA Motheo

Limpopo: 5 YDF Instructors
Current & former partners: African Youth Development Fund (AYDF), Mokopane Academy of Sport

Mpumalanga: 4 YDF Instructors
Current & former partners: Scouts Association South Africa (SASA), Seabe Sports Development Association (SEABE), Siyakhula Youth Health and Development Organisation

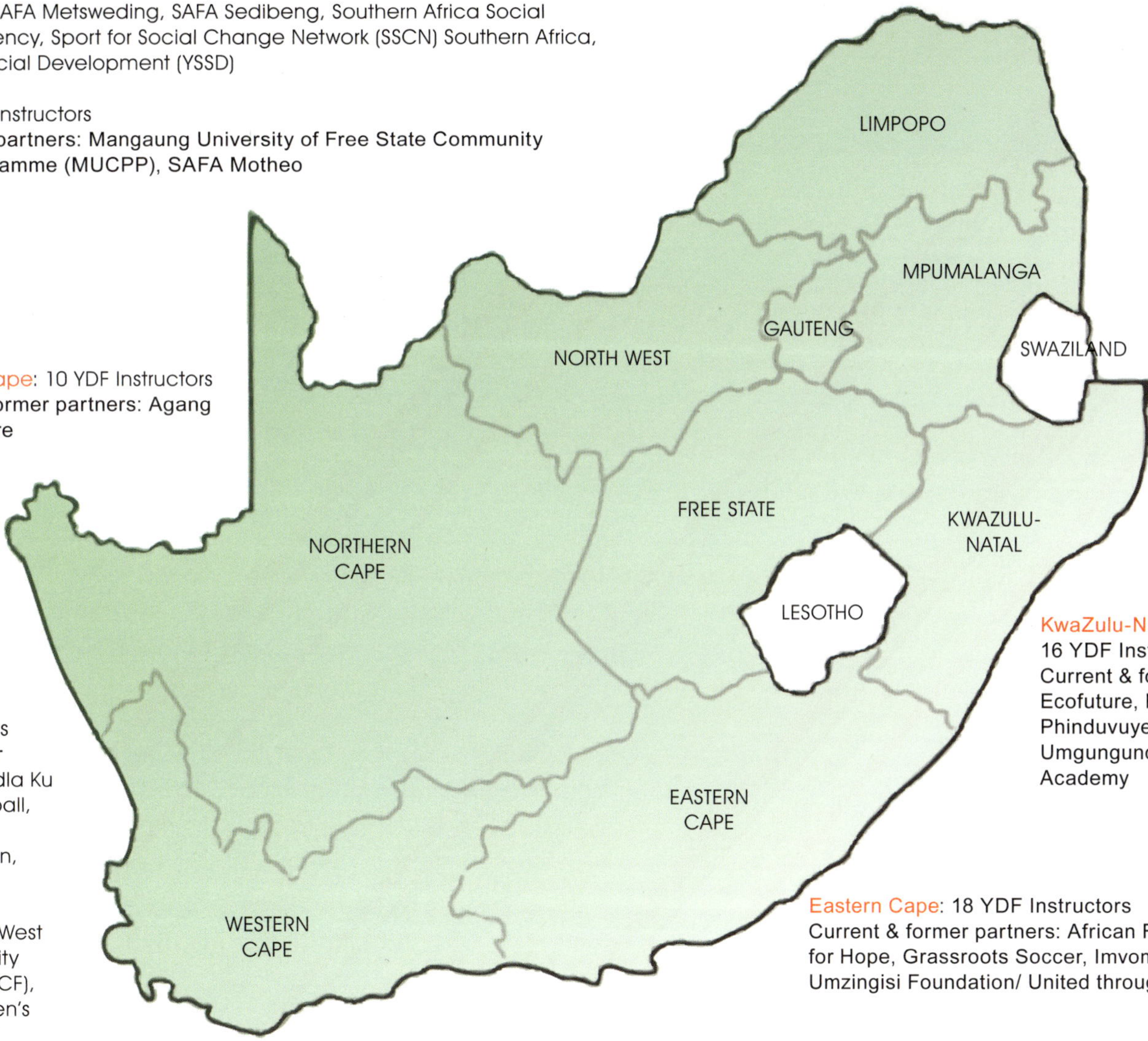

Northern Cape: 10 YDF Instructors
Current & former partners: Agang Youth Centre

Western Cape:
22 YDF Instructors
Current & former partners: Amandla Ku Lutsha/ Edufootball, Arise, AYEP, New World Foundation, Oasis – Reach for your dreams, Soccer 4 Hope, West Coast Community Foundation (WCCF), Yabonga Children's Project, Inyanda Yabatsha

KwaZulu-Natal:
16 YDF Instructors
Current & former partners: Ecofuture, Inxuluma, Phinduvuye, Sportstec, Umgungundlovu Sport Academy

Eastern Cape: 18 YDF Instructors
Current & former partners: African Footprint for Hope, Grassroots Soccer, Imvomvo, Umzingisi Foundation/ United through Sport

Foreword, UNOSDP

Wilfried Lemke, Special Advisor to the United Nations Secretary-General on Sport for Development and Peace, UNOSDP

Recent history has plainly exposed the importance of strengthening participation and inclusion of young citizens into all aspects of society. Fostering child and youth development is essential in order to create responsible citizens, achieve the Millennium Development Goals (MDGs) and promote peace.

Recognition of sport and physical activity as an effective tool for development and peace is growing throughout the world. They are specifically mentioned as such in an impressive and ever more growing number of policy and programme documents by international organisations – including the United Nations – as well as governmental institutions.

Sport becomes especially relevant in the realm of promoting child and youth development, and more so when it touches the lives of vulnerable children and young citizens.

The Youth Development through Football (YDF) project of the German Development Cooperation (GIZ) bridges the gap to those most vulnerable sections of the population, to children and youth for whom access to education, sanitation and health information is not a given, to those surrounded by violence, disease, environmental pollution and more social challenges. It uses the enthusiasm for sport in general – and football in particular – to equip youth with life skills, increase knowledge on HIV and AIDS, encourage peaceful conflict resolution, advance gender equality and promote environmental awareness.

I have myself been exposed on many occasions to the work carried out by GIZ through the YDF initiative, and have visited several of the projects, namely in South Africa, Kenya and Rwanda. Hence I can confidently attest to the high standard of work that has been carried out by GIZ and their partners, and to the strong impact that they have on the ground.

As a matter of fact, the YDF methodology is at the cornerstone of a new youth development project which we are supporting in Ukraine through a financial grant. By adapting the methodology to the Ukrainian context and receiving the support of GIZ, the project aims at leveraging football and football-related training exercises as a means to empower disadvantaged youth across the country and inspire them to become themselves agents of change for their communities.

Stories from the Field provides us with an exciting insight into this innovative approach and on the impact of YDF on individual beneficiaries. This extensive collection of case studies is a very valuable publication. It gives us even more strength to pursue our path and to continue promoting and using sport to change the lives of young boys and girls worldwide.

Wilfried Lemke visits Soccer 4 Hope in the Gugulethu township of Cape Town, South Africa, 6 July 2010. The organisation is a member of the YDF network.

SPORT FOR DEVELOPMENT AND PEACE

UNITED NATIONS
UNOSDP
Office on Sport for Development and Peace

Foreword, SRSA

Fikile Mbalula, Minister of Sport and Recreation South Africa (SRSA)

If one has the privilege to travel to parts of South Africa, one can behold breathtaking scenery and vibrant people forging their path for a brighter and prosperous future.

A picturesque and wondrous sight in South Africa is the Boabab tree. Prime specimens of this magnificent tree can be found in Modjadjiskloof in the Limpopo Province. Africa is symbolised by these magnificent trees. Many local people believe that the Baobab tree grows upside down due to the massive trunk, which gives rise to thick, tapering branches resembling a root system. There is a tale amongst indigenous people, which tells of how God planted the trees upside down.

A botanist will tell you that even a impressive tree the size of the Boabab starts with a seed pod that can nestle in a man's hand. The Baobab tree requires a full-sun location with plenty of room to grow and can be found in arid areas.

This may not be a botany book but beautiful symbolism can be found in this tree for sport in South Africa and its people who play. Even after the entire tree is cut down, it simply resprouts from the root and continues to grow. Sport was blighted and oppressed during apartheid and opportunities where not afforded to all citizens to compete, with their talents figuratively being cut down. At the dawn of a new South Africa, sport can resprout and begin to regrow.

Even in the remotest areas within South Africa, the Boabab requires certain conditions for growth to occur. Water from the heavens provides the much-needed resource for growth to occur. The growth of sport in South Africa has been supplemented by resources and funding from the YDF project.

I conclude with a myth that it was once believed that Baobabs were in danger of becoming extinct. This was before botanists realised that the small trees do not resemble the mature trees at all. So, with this, my hope is that when you return to the fertile land of South Africa you no longer recognise the sport that you left behind, but a majestic and thriving thing you helped to grow.

On behalf of the South African Government, I want to convey my sincere appreciation to GIZ, the European Union and the German Government for their contributions and cooperation to grow sport in South Africa.

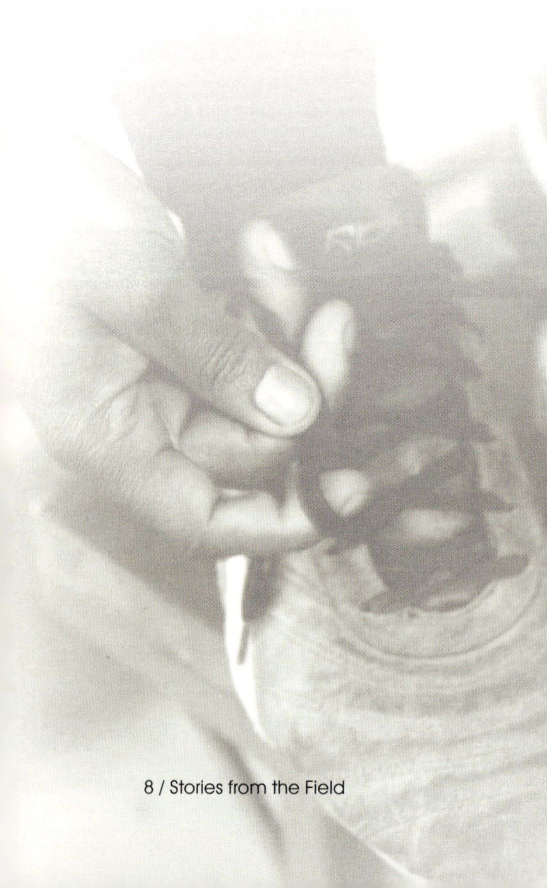

sport & recreation

Department:
Sport and Recreation South Africa
REPUBLIC OF SOUTH AFRICA

Foreword, BMZ

Dirk Niebel, German Federal Minister for Economic Cooperation and Development, BMZ

As a member of the G-20 and an emerging economy that – due to its potential – recently joined the BRIC states Brazil, Russia, India and China, South Africa is of significant importance for the political, economic and social stability of its region. By hosting the first FIFA World Cup™ on African soil in 2010 the country gained worldwide attention and recognition, and contributed to the rising reputation of the continent. Germany was able to help its partner country, South Africa, prepare for this stunning sporting event, sharing lessons learned from its own recent experience as host of the FIFA World Cup™ in 2006.

The German Federal Ministry for Economic Cooperation and Development cooperates with the South African Government in order to fight poverty and social inequality, secure stability and jointly protect global public goods. South African–German development cooperation accordingly focuses on energy and climate. For example, by creating "green" jobs, on good governance – including the prevention of social violence – and finally, on HIV prevention.

Against this background, and with the 2010 World Cup on the horizon, we started supporting the programme "Youth Development through Football" (YDF) in 2007, focusing on those who are most vulnerable: disadvantaged children and adolescents from poor rural areas or townships. Since then, YDF has successfully offered capacity development for coaches, teachers and peer-educators, enabling them to take advantage of the enthusiasm of young (South) Africans for football. Through sport, YDF enhances "life skills" like team spirit, fair play and non-violent conflict resolution. Education on HIV and AIDS, violence prevention, environmental awareness and promoting gender equality is spread on the soccer field – in short: YDF's goal is to ensure that the World Cup leaves a durable and sustainable legacy in South Africa and its neighbouring countries. YDF has so far reached over 50,000 children and young people, hopefully offering them better prospects for their future lives.

YDF's educational "sport-for-development approach" is being replicated and adapted in other countries and continents. The programme's partnerships with the European Union and two prominent companies, UNESCO's patronage for the Toolkits developed by YDF and, last but not least, the very convincing results all point to the lasting positive impact of this programme.

BMZ Bundesministerium für wirtschaftliche Zusammenarbeit und Entwicklung

Foreword, EU

Andris Piebalgs, Commissioner for Development, EU

Sport has the power to unite people across different cultures and backgrounds, as was clearly illustrated in South Africa by the 2010 FIFA World Cup™. It also offers a wonderful opportunity to imbue both players and followers with a sense of responsible citizenship through fair play, teamwork and leadership. At the same time, sport has an important role to play in development and I'm glad that the European Union actively participates in related projects. The EU supports, for instance, the successful South African-German Youth Development through Football (YDF) partnership with a contribution of some 6 million euros (R60 million).

During my recent visit to South Africa, I was able to see for myself the difference that sport can make in people's lives. There, the Youth Empowerment Programme, helps youth development through local communities, for example by training sport assistants – contract workers between 18 and 25 years old who were often previously unemployed – to be YDF Coaches.

Another example comes from Mahlasedi Masana Primary School, which I visited whilst there. A street-soccer pitch has recently been built by the Deutsche Gesellschaft für Internationale Zusammenarbeit (GIZ) GmbH, with funds also from the European Union.

The programme was set up to improve the life skills of young people and help them to find jobs. It is already achieving results. Overall, more than 50,000 youths between 12 and 25 years old will be supported in South Africa, of which at least 40% will be girls.

Through sport projects we can convey different messages on education, gender and health but also peace and mutual understanding. I'm determined to continue our social investments and in my latest policy proposal, known as the "Agenda for Change", I pointed out that a minimum of 20% of EU aid will go to human development, including support for youth through sport, culture and education. I believe that continued youth development is vital to a society's stability, sustainability, growth and well-being, since it may prevent early school-leaving and promotes active citizenship.

I wish to take this opportunity to thank all our partners, but specifically the Department of Sport and Recreation South Africa (SRSA) as well as GIZ, for the hard work and dedication that has turned a modest programme seeking to integrate girls and boys from different backgrounds and ethnic groups, into a towering example of how commitment and cooperation can build a lasting legacy for a country, a continent, and a partnership for progress.

Foreword, GIZ

Andreas Proksch, Director General, Africa Department, GIZ

After the transition from apartheid to democracy, Germany and South Africa soon began their bilateral cooperation, and in 1996 the Deutsche Gesellschaft für Technische Zusammenarbeit (GTZ, called GIZ since 2011) established a country office in Pretoria. GIZ supports the country mainly in the areas of democratic development and good governance, HIV/AIDS, as well as climate change and energy. In addition, GIZ implements various ambitious programmes in skills and youth development.

The Youth Development through Football (YDF) is a crucial pillar in the framework of German–South African cooperation. The project has elicited much publicity before, during and beyond the FIFA World Cup 2010™, attracting the interest of politicians and international organisations alike. The experiences of this programme – how to use football for developmental goals – are already utilised in other programmes and will benefit youth development programmes all over the world.

The impact of the programme at the beneficiaries' level is well documented in numerous studies and impact assessments carried out by the project and the University of Johannesburg. Nonetheless, the present case studies complement the research and represent a valuable certificate of the impact on the personal lives of individuals. They give us a feeling of what it really means to use sport to attract, reach out and impact on the lives of young boys and girls.

giz Deutsche Gesellschaft für Internationale Zusammenarbeit (GIZ) GmbH

Introduction

by Professor Cora Burnett, University of Johannesburg

When we think about Africa, we often recall pictures of a wild countryside, rivers, forests, deserts, buzzing markets and media stories of poverty, devastation and hunger – a prime continent where branches of human justice are still thin and often absent. Those of us who have made Africa our destination are witness to a different experience as we meet people who charter a life of resilience and construct a social world of their own. Extended family ties stretch far and deep and so do family feuds, political turmoil and the many manifestations of poverty – painting a human canvas on which development programmes (including sport-for-development) need to blend in strokes for reciprocal change.

The GIZ/YDF programme is possibly one of the most significant programmes in terms of philosophy, methodology and scope that sheds light on development work as it is embraced, transformed and assimilated into real-life experiences. It is at this embedded level of social uptake that the selected 15 stories in the text and the detailed 45 stories on the CD articulate real-life experiences as told by programme participants, peer-educators and YDF partners who took up the responsibility and challenge to change lives for the better. Multiple voices give an expression to the meaning and dynamics of sports-related interventions as they affect individuals within a myriad of social relationships and contexts. Due to the sensitivity of information, pseudonyms have been used and informed consent was received from all participants.

The uniqueness of this publication is that it not only captures all these voices to constitute stories from nine African countries, but validates a ripple of impact where context is a prism for seeing and knowing. How do real people see and benefit from a sport-for-development programme? How does the North meet the South in partnership and trust?

After reading these stories, you will not merely see young boys or girls kicking a ball and appreciate the novel aesthetics, but you will smell the dust and know it is a family and community at play. You will realise that aspirations and expectations are born from a need for entitlement in a field where the pitch is only the beginning of a journey. May this brief encounter do justice to all who shared their lives with such honesty and hope.

UNIVERSITY OF JOHANNESBURG

The YDF Project

Sport is far more than mere competition, it's more than simply spare time and it's more than just entertainment for the spectators. While its intrinsic values, such as fairness, mutual respect, and healthy lifestyle have always been known, the potential that sport can unleash when systematically included in processes of social change has only recently experienced a rise in recognition.

More than any other sport, football exerts an exceptional power on the people, more so on the youngest ones. While Germany experienced the positive power of football by hosting a memorable FIFA World Cup™ 2006, the prospect of the first FIFA World Cup™ ever to be staged on the African continent filled many Africans with great pride. It is against this background, that the idea of "Youth Development through Football" (YDF) was born.

The Youth Development through Football (YDF) project started in 2007 as a legacy of the 2006 FIFA World Cup™ in Germany, envisaged to contribute to the legacy of the 2010 FIFA World Cup™ in South Africa. The project uses the popularity of football to promote youth development, especially of disadvantaged girls and boys. The project's major objective is to establish and support youth development initiatives in selected African countries where football training is combined with non-formal education measures on life skills and relevant development issues. The project is funded by the German Government through the Federal Ministry for Economic Cooperation and Development (BMZ) and co-funded by the European Union (EU) with a total of 16 million euros. YDF is implemented by the Deutsche Gesellschaft für Internationale Zusammenarbeit (GIZ) GmbH, in partnership with the Department of Sport and Recreation South Africa (SRSA).

Additionally, YDF entered into two public–private partnerships with prominent companies: the partnership with Volkswagen South Africa supports HIV/AIDS prevention in the Eastern Cape, while the cooperation with Nike South Africa aims at strengthening the Sport for Social Change Network (SSCN) Southern Africa, offering support to its secretariat as well as capacity development for SSCN member organisations.

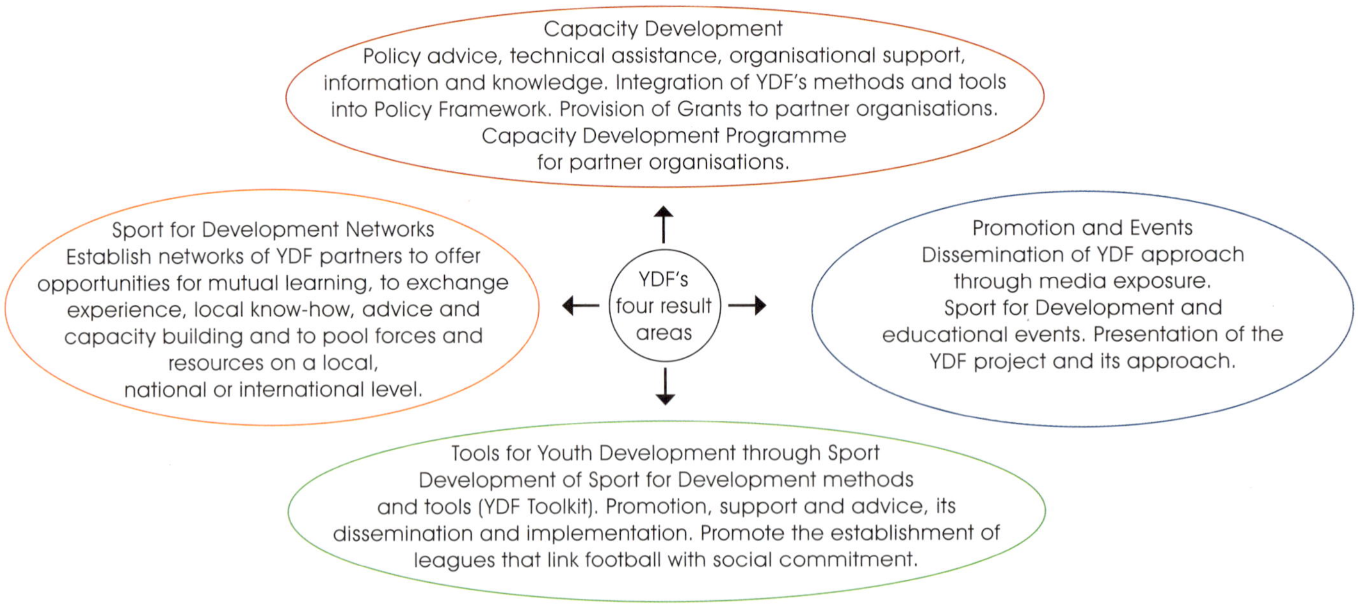

The development and promotion of tools and methods of youth development through football represent the most crucial cornerstone of YDF. The YDF Toolkit was developed in close cooperation with experts from the specific fields targeted in the YDF manuals. The foundation of the YDF Toolkit is formed by the generic YDF Manual for Instructors and the YDF Manual for Coaches. The two manuals on teaching football and life skills uncover the possibilities of integrating non-formal life-skills education in football training sessions, thereby creating an innovative and fun learning approach on the field. The YDF Manuals for Violence Prevention, HIV/AIDS Prevention, Gender Awareness, Disability Inclusion and Environmental Awareness, form the set of short modules that deal with specific social development topics in a more detailed way. In 2011 the United Nations Educational, Scientific and Cultural Organisation (UNESCO) granted its patronage over the YDF manuals, prominently emphasising the importance and the potential of sport for youth development and education. The YDF Manual for Sports Event Management serves as added value to the YDF Toolkit, offering organisations a handbook on how to organise events with a strong focus on sport-for-development approaches and activities.

The YDF Toolkit, under UNESCO patronage since 2011.

Through numerous partnerships with government institutions, non-governmental organisations and football associations as well as other GIZ projects and offices in Africa, YDF is reaching out to more than 120,000 children and youth in 10 African countries (see map on page 4). Additionally, ahead of the UEFA Euro 2012, GIZ established a project based on the YDF approach and methods in Ukraine, focusing primarily on HIV/AIDS prevention among youth.

After having concluded its first project phase, YDF will continue its work until December 2013, concentrating first and foremost on strengthening the quality of YDF Instructor and YDF Coach trainings and youth activities, as well as on ensuring sustainability of the YDF approach within its partner organisations from grass roots up to national and international levels.

Stories

'But when there is a soccer game, the picture changes: the grounds are full and the shebeens stay empty'

Mavis, Dona's Mates, South Africa

Mavis is a coach for Dona's Mates in Orange Farm. The 29-year-old grew up with her eight siblings on a farm in Fochville, but moved to the township of Orange Farm at the age of thirteen. Today, Mavis coaches a children's team and a ladies' team for Dona's Mates and lives with her two children at her sister's home.

Background

Mavis's childhood in Fochville was peaceful. She lived on a farm with her parents, her five sisters and her three brothers. But this atmosphere was suddenly destroyed by a fight between Mavis's father and the owner of the farm. As a consequence, the family was forced to leave and eventually settled in the township of Orange Farm, 50 kilometres south of Johannesburg. The family's move was difficult as they could not practise subsistence farming and thus provide food and basic necessities for all. Not only was the urban life a big change for them, they could also not find affordable accommodation for the family to live together. A splintered life began for most of them and Mavis stayed with her sister who was living in Orange Farm with her boyfriend.

Mavis had both pleasant and harsh experiences during her primary school

years, but was lucky to have more caring teachers and positive experiences during her high school years. Besides sports and music, she enjoyed debating. 'I liked to argue about things and that you can disagree with people,' she remembers. But during school years, Mavis was often exposed to roughness and violence. 'Our teacher was very strict and he would hit us a lot. If there was some candlewax on your book or the book was slightly burned, he would hit you on the hands. The principal would also hit you if you were late for school. He would wait for you at the gate and you would be in big trouble,' she recalls.

Community life

As poverty is rife among most of the residents of Orange Farm, the lack of resources and how it affects the lives of individuals and collectives, as well as the whole community, is not being done justice in a brief synopsis. However, the main issues that negatively impact on the quality of life in the community provide a mere glimpse of the local context.

Mavis's sister Sharleen mentions the high number of house break-ins in Orange Farm. 'But we've come up with a good system in our neighbourhood,' she explains. 'If someone sees someone else doing something illegal, he or she whistles so everyone comes out and helps to catch the culprit and to call the police.' This has contributed to lowering the crime rate. Orange Farm has to cope with a very high level of unemployment. This leads to the fact that many people cannot afford food or health treatment. Sharleen points out that 'the youth, also due to lack of perspective, is increasingly affected by alcohol. But when there is a soccer game, the picture changes: the grounds are full and the shebeens stay empty. The 5-a-side is busy and on holidays people go to watch or play with umbrellas and cooler bags; they don't go to the shebeens, especially the boys at the pitch'.

A much more positive view is expressed by Mavis's "husband" Petrus – he is the father of her two children, but they cannot marry because there is no room to live, or money to pay *lobola* in addition to their "other" families who are dependent on their contributions. He said it is a wonderful community to live in because of the people who are friendly and helpful. He doesn't like the negative perception of the community, as 'good things are happening here too'. Petrus is also a coach for Dona's Mates and has been with the "club" from the beginning.

The YDF programme

Although Mavis is earning a stipend as a 'regular salary for the first time in her life' and has to look after her two children, she also has to contribute towards her sister's household. Her "husband" can hardly

afford to assist her financially, as he is also responsible for looking after his father, unemployed brothers and an extended family. Again, Mavis's sister Sharleen explains the dynamics in her home and the need for money:

> 'Mavis is so kind. She gives me all the grant money and most of her stipend that she is earning at Dona's Mates. When they have tournaments, I help with the catering. With the leftovers, we can still eat for two days.
>
> Because of Mavis' stipend, we could paint the house. We are putting money away to make a bedroom for her and her children in the shelter. When it is done, Mavis's "husband" can come and stay with us. The family can then be together. For now, she has been sleeping in the kitchen with her children.'

Mavis has a long association with Dona's Mates that began during her secondary school years. She would regularly go and watch soccer games and was then invited by Petrus to get involved. She assisted them by washing the jerseys of players. Two years after she left school, she had a baby boy and later her daughter was born. In 2009, she and her sister were "down and out" and her sister found a job as a domestic in Triomf for 18 months. As she was earning only little and had to sleep in, she could not cope alone at home. She could also not meet the cost of the crèche on grant money, therefore the stipend from Dona's Mates at least 'put half a loaf of bread on the table'. The personal satisfaction, recognition and acceptance as a coach are what keep Mavis going. She tells about the following rewarding experiences:

> 'I love my job as a coach and like to work with the Under-10 players and with the ladies. The small kids make you happy every day. They can make you shout and make you laugh. I even forget that I sometimes don't have food.
>
> The ladies take me as their sister. They trust me and I help them to tolerate life. The young ones also like to compete, but to be a coach for the ladies is challenging. They have many moods. Imagine 15 different women all together.

One lady is ashamed of her big boobs and we need to give her confidence. Others do not have shoes and sometimes you see them very down, and you know it is bad just by reading their background.'

Mavis is making the most of her situation. She realises that it is not easy to change things in the community as they do not have the support. There is a lot of poverty and crime. The criminals will even steal and cut the side-fence stands for scrap. Mavis explains how coaching has enriched her life and how she feels it is her responsibility to try and make a difference:

'I will not move, despite these negative aspects. Orange Farm is home and it is less expensive than Soweto. We are all poor, even the manager. He supplies us with words and makes us strong. We can coach and make life better for others. We have seen the changes with our players. In the past they would complain and fight – now we partner them and they communicate much better.'

DONA'S MATES YOUTH
DEVELOPMENT ACADE

'The benefit of teaching life skills is that I also changed. If I can make a difference in the life of just one child, I will be happy'

Mamello, Umzingisi Foundation, South Africa

Mamello is a peer-educator and coach for the Umzingisi Foundation, where he combines life skills education and football training. The 29-year-old grew up in Port Elizabeth's township of New Brighton and is still living there with his three uncles, his grandmother, his aunt and three cousins. His mother only stays over at weekends as she is a domestic worker and has a live-in job.

Background

Several of Mamello's immediate family members have passed away due to AIDS, a disease which is still affecting the family profoundly. His eldest uncle – initially the main caretaker of the household – tells a story of poverty, disease and a belief system that provides them comfort, as much as it serves as a tool of how to deal with all these "calamities".

> 'Since my younger sister died of AIDS, the disease has been in the family. Both her children also died from AIDS, and also my twin brother. It is as if the devil wants to destroy our home. I am also HIV positive and am getting a grant for it. I do not know how this could happen.

My skin was burning and I decided to pray and go to church to be cleaned. As I wanted to go to church and went for a drink of water to the tap in the yard, a woman stopped to talk with me. She saw snails on the ground and said that the devil has willed the demons to enter my body. I went inside and looked in the mirror. I saw that I was cursed and that my face had purple patches where the snails entered. Now I am going to the pastor to pray for me.'

In addition to the "terrible infliction", the family is mostly dependent on welfare grants, the eldest uncle's AIDS grant, Mamello's contribution and that of his mother. Mamello dropped out of high school, but finished his education at night school, as his best friend took care of him and encouraged him to finish his matric.

In the past 11 years, life has not been easy as Mamello was not able to find any kind of steady employment, except to earn some "contract money" or a stipend – first as a volunteer with the NGO Grassroot Soccer and then with Umzingisi. Mamello's greatest wish is to become economically independent and stay on his own, especially since he is going to be a father in the near future. 'But I don't have the money and I can't find other work. However, I find my current job very valuable as I can make a difference in the lives of children.'

'I am to be a father soon and I had some relationship issues. Now it is better and my girlfriend will stay with her parents and I will go and see my family as often as possible. I want to be part of her life, but cannot do so much with R2,500 per month [245 euros / US$325]. I have to pay *lobola* of about R20,000 [2,000 euros / US$2,600] to R30,000 (3,000 euros / US$3900]. It will be a long time for me to save this amount of money. I will pay this in instalments and have to feel the pressure on my own.'

The additional financial burden of *lobola* (i.e. the payment of a "price for the bride" in Xhosa tradition as to replace the woman's labour by acquiring another) adds to fragmentation of family life for young parents.

Community life

New Brighton is an impoverished township, but it poses a vibrant sporting culture as many sports are offered in the community. Many local sport clubs offer regular activities and weekend matches for the community.

Problems in the community are mostly associated with youth drinking out of control and committing crimes. They often 'hang

out around shebeens and rob people at night'. According to Mamello, these youths will not necessarily join their sport activities. He sees as the key solution to address this issue recruiting the youth and to 'integrate them into the church activities, as well as in sport educational programmes'.

The YDF programme

Mamello has an active sporting background. In primary school, he also took part in judo, rugby, cricket and soccer. Later, his friend who was a manager at Grassroot Soccer recruited him as a coach, for which he would earn some income. It is with this organisation that his journey as a "life skills coach" started:

> 'I started to take life more seriously when I became a coach. I became a role model and was very caring. I found inner peace and then I started reading motivational books like David Molapo's *I Can Foundation*.
>
> I wanted to go back to school to be able to study further. My uncles had been disappointed when I left school because they had put their faith in me for studying. I never wanted to disappoint anybody again.'

This attitude of care spilled over in Mamello's current involvement with Umzingisi and as a peer-educator in the YDF programme. His stepsister bears witness to his increasing maturity in terms of him 'applying his knowledge to teach me to take responsibility for my choices. Mamello always gives me good advice. A weight gets lifted when I talk with him, as he is our role model in the family. Mamello is ambitious and will make it in life'.

Mamello sees his current work as a calling and expresses his commitment to answer in the needs of his "constituency" in relatively philosophical terms:

> 'What I like most about my current job is that I can give spiritual support. I am resilient and can teach the children to be the same. I have learnt so much and can also teach them some new tricks of life.
>
> The benefits of teaching life skills is that I also changed.

If I can make a difference in the life of just one child, I will be happy. One of the older children phones me often. I changed his life and took him to the church with me. Christianity is a lifestyle.

Some kids do not have a father and I try to assist them as a male role model. Another friend is an alcoholic, but now I am looking after him. His friend's brother participates in Umzingisi's programme "School of Excellence" and I try to keep them all out of trouble.'

He is keenly aware of teenage girls whom, he feels, need to be warned against having sexual relations with boys. 'I always remind the girls of the "Seven Bs": Books before boys because boys bring babies.'

UMZINGISI

'We develop the personalities of coaches and peer-educators. The participants acknowledge and listen to us. They come to us with their problems and they trust us'

Nkosozana, Soccer 4 Hope, South Africa

Nkosozana is a coach for the NGO Soccer 4 Hope, where she has advanced to a management position. Though born in the South African province of the Eastern Cape, the 23-year-old grew up in one of the biggest townships of Cape Town, Khayelitsha. The Xhosa name of the settlement fits with Nkosozana's story as well as with the background of so many of its other residents that have been arriving there over the last few decades: it means "new home". Today, Nkosozana is still living there with her mother, her younger sister and her cousin.

While Nkosozana has had a troubled relationship with her father, she explains how her mother was always looking out for her safety in Khayalitsha:

> 'I came to live here in Khayalitsha when I was two years old. I was the only girl and was always surrounded by boys. The primary school closest to home was a 45-minute-walk away. It was not safe to go there alone. Once I was harassed coming back from school so I got an "older brother" James to walk with me and protect me.'

Nkosozana said she was very shy in primary school, mostly because she was

a 'Xhosa in a Coloured school', but soon she adapted and was chosen as a prefect. In Grade 7, she became very quiet because of an abusive relationship with her father.

> 'My father was abusing my mother and he was hitting her so that she had some miscarriages. I felt I could not speak to anyone and felt sidelined. In 1999 I was sent away to live with my aunt and only got back on Sundays. My little sister was born but my father would not stop. One day he was hitting me and my mother when my little sister came into the room, asking us why he was hitting us. I felt so very ashamed as I did not want her to see this. I started hitting back at my father and took my sister away.
>
> During this time, I stayed mostly in my room, hiding away, as I was not comfortable in any other room. It went on and on, and my father only left us when I was in Grade 11. He moved everything out of the house. Only my bed and a few blankets remained.'

For Nkosozana, her father's abuse and desertion destroyed her chances of further studies and left emotional scars. She had to grow up fast and assist in taking care of her family.

> 'In Grade 11 I got very poor marks, but I managed to pass Grade 12. After that, I went to the Oval International College to study business management. But after six months I fell ill and I didn't have money to continue as my uncle could not afford it anymore. My mother was selling sweets to send me to school, but I didn't want to take her last money. I just wanted to find a job.'

Having worked at Soccer 4 Hope for two years as a volunteer and then earning a stipend of R2,000 (200 euros / US$260) was a "godsend", and Nkosozana relates her excitement about being able to contribute money to the household:

> 'As a volunteer I did not earn anything. I was volunteering for six months, but was also looking for a job. Finally, in March 2008 I got a job at Soccer 4 Hope. I was so excited. At the end of that month I opened a bank account to put in my first salary. Tradition would have it that I gave the money to my mother.'

During all this time, Nkosozana's life was in emotional turmoil and she wanted to share her story with others. She wrote poetry and was

chosen to tell her story on national television. Not only did she find some kind of emotional relief, but she mattered and she had a voice.

The YDF programme

Being respected as an "All Star" coach and life skills manager for Soccer 4 Hope provided Nkosozana with coping mechanisms in the otherwise tormented life. The organisation became a safe haven for Nkosozana, a constant in the otherwise hectic social world. She explains how she got involved and her journey in this organisation:

> 'In 2006, my uncle who was a coach at Soccer 4 Hope, recruited me also as a coach. At first I didn't like working with kids, rather I wanted to hit them.
>
> Yet I stayed as a volunteer and soon I started appreciating it. I had a sense of belonging at work and a sense of home. I got fully involved in the community through the girls I was coaching. I wanted to plant the seed in their lives so that they could face the challenges. I wanted to be that someone for them. I wanted for them like it was for me – to move into that other zone, to cope.'

Later, Nkosozana was trained as a YDF Instructor, so that she could train other coaches. One man, whom she had trained in 2008, took over the "All Star" coaches and became the manager of the programme. Nkosozana was happy that he could support her. 'I only knew groundwork and could not move into a management position. He had training at a university in sport management and was also experienced in managing an event.'

Today, Nkosozana loves working with children and finds such satisfaction to 'share her life' and having a meaningful engagement with young people:

> 'At Soccer 4 Hope, we develop the personalities of coaches and peer-educators. The participants acknowledge us and listen to us. They come to us with their problems and then trust us – we find common ground as the coaches are

also from the community. Some kids do not want to go home after practice; there is too much abuse at home and then they find they cool down during our activities.'

Being involved with the YDF programme and the opportunities it opened up for her to travel abroad, became the highlight of her "career". She is in awe when she recalls the blessings she has had by meeting all the wonderful people. By being a spokesperson for her community, her organisation and the YDF programme gives her a platform where she can come to her own. At some international forums, she is representing the 'youth of South Africa' – an ambassador's role that is very dear to her. She sums it up in one sentence: 'Soccer 4 Hope brings balance to my life.'

5

'The peer-educators are like our older sisters and brothers. They guide us and are very open and friendly'

Kgabiso, SEDYEL, Botswana

Kgabiso is a 14-year-old girl from Ramotswa, a town situated 25 kilometres south of Gaborone, the capital of Botswana. The Form 2 student plays soccer in the South East District Youth Empowerment League (SEDYEL), a youth-initiated football league which uses sport as a tool for community development and for building youth leadership skills. Kgabiso lives with her parents, two older sisters and one little brother.

Background

Kgabiso's father works for Water Utilities and the 19-year-old sister is working as a secretary. Her 24-year-old sister is not working as she has a little baby and has to stay home to look after him. Her mother is a businesswoman, selling chickens and eggs. Most of the income for the household is generated through this small enterprise and also pays for the youngest boy's private education at an English-medium school. Kgabiso says that her brother should be educated because 'he is a clever boy and serious about studies. She helps her mother so that she can make money for her brother's schooling.

Community and social life

Community life seems to have its pros and cons. For more than five years, the chief of the village has been a woman 'who is respected by everyone'. For Kgabiso's mother, the situation of traditional leadership in the village as a positive aspect. Volunteerism and active citizenship are highly valued in the community. 'We have home-based volunteers in the community who care for people with AIDS, tuberculosis and diabetes. They visit these people and give them food – they are assisted by social workers and nurses,' explains Kgabiso's mother.

However, there are many negative influences in the community as well – the behaviour of often idle and unemployed youth is a main issue. Kgabiso's sister is worried about the youth:

> 'The youths do not respect older people. They go to school, but when they go out, they want to do what they want. To me, Kgabiso is one of these "problem children". She goes with friends who have a bad influence on her. At least soccer keeps her busy and out of trouble and bad company most of the time.
>
> There are many school dropouts, especially from junior school and a lot of pregnancies. There is a shortage of job opportunities – many graduates are without jobs. This in turn contributes to teenage pregnancies which cause lots of problems for families. The naughty ones are mainly older boys and younger girls. The boys who dropped out go for the young girls and then they become totally out of control.'

For Kgabiso, community life is mostly interwoven with her relationships with her peers – with her close female friends and with boys. She also reflects on various relationships with boys and with peer-educators from SEDYEL, explaining:

> 'In the community guys do not treat girls in a good way. They will just tell you that they love you, but if you do not want to go with them, they will beat you. One of the boys used to treat me like this, then he joined the soccer team and at New Year's Eve he asked me to forgive him. I forgave him and now he is talking nicely to us girls.
>
> If boys are very rude and fight a lot, the peer-educators will talk to the parents and they will discipline them. The peer-educators are like our older sisters and brothers. They guide us and are very open and friendly.'

Kgabiso was very candid about her less favourable characteristics and admitted to bouts of drinking and "mixing with the wrong crowd" that contributed to discontent with family members and unacceptable social behaviour.

> 'I am not always a good girl. During school holidays, I will go out with older girls and drink over the festive season. We will even drink hot stuff like brandy and Autumn Harvest, a sweet wine. We go to many parties and we will be served drinks as they do not bother to ask us our age.
>
> I know this is not working out for me, and since I have been going to SEDYEL, I have made different friends. It is just over the long holidays that these other girls come and ask me to go with them. It is difficult not to. I am still working on changing, but it's not so easy. Maybe if I will become a peer-educator I will behave better.'

The YDF programme and sport participation

Kgabiso played football in primary school, but it was not serious and mainly an activity in which she took part with some of her friends. She was never good at academic subjects and preferred to play sport. She failed Form 2, despite having passed well in primary school. However, she maintained her sporting success in high school where she took part in football, softball and shot put. But the school coach did not have much passion for the job and was lacking in proper communication skills. Kgabiso recalls. 'He was always shouting a lot. Also, it was not widely accepted that we play football in secondary school and so we did not have a full team. Therefore I joined SEDYEL for football.'

'I have been playing now for two years at centre and like very much how they organise games and have life skill talks. We have learnt so much – how to play, how to communicate and how to treat people. We also learnt how to protect ourselves from not getting infected with HIV.

We talk about diseases to our friends and share information. I have about five close friends and share with them what I learn from the programme. I talk to them about HIV issues and how always to protect yourself against becoming infected. We should avoid sexual relationships and drug abuse. First they thought: who is she to tell us such things? I cannot stop them doing these things, but I made them "brave" to go to the clinic and get protection.'

BOTSWANA
PIONEER
LOG HOMES

PRO★STAR

David, Vijana Amani Pamoja, Kenya

David is a peer-educator and coach for Vijana Amani Pamoja (VAP), an NGO based in Nairobi that educates youth by using football. The 26-year-old used to be a very good football player, but then he was injured and 'things did not work out' for him. He is also the father of a young daughter and is currently living on his own in a rented room in close proximity to the VAP offices in Kiambiu, the second biggest slum in Nairobi.

Background

Together with his parents, his uncle and his three younger sisters, David grew up in California, an area of Nairobi close to the slums. While struggling academically at school, he performed very well outside the classroom, playing soccer regularly and competing in athletics. But after his father lost his job, the family had to move to Kiambiu, one of Nairobi's slums situated only four kilometres from the city centre.

David's life took another turn when he was sent to boarding school. His mother explained that he was 'failing at school and we thought it best to send him away to get his mind off just playing football with his friends. He was a good boy but always playing. He even kicked the ball inside our house'.

'I get the opportunity to change the lives of my players. I give them good advice, even with the little that I know'

David is aware of the reason for him being sent to boarding school, but found it a trying time in his life. The school was in the province of Nyanza, a 12-hour bus ride from home. David did not enjoy his time there, as the conditions were very harsh and violent. But soccer remained his passion:

> 'I became the school's soccer captain in the fourth year. The teachers liked me as I put the school's name high, so they hit me less for my mistakes.
>
> My friends and I attended regular practices, which made me not to indulge in drugs. We players loved each other. We would scratch each other's backs and we could open up to the coach. He was not as strict as the other teachers or our parents. He talked to us in our own language and we felt he was part of us.'

After school, David had to find his feet, studying any "affordable" course and trying his utmost to find employment. He worked in markets, selling second-hand clothes and chickens, but was not successful. His life changed when he met his girlfriend:

> 'After being together for five months, we had unprotected sex and she became pregnant. I was only 22 years old. We were young, didn't have work and were not prepared for the responsibility of having a child. After the birth of our daughter, I tried to provide for my girlfriend and the baby, but it was not enough. She is now engaged to somebody else who earns enough money. Sometimes my daughter tells me to wait while he is playing with her and that hurts a lot. He is now going to be her other father, just because of my financial status.'

Community life

David's younger brother said that because of the high level of crime in the community, there should be an improvement in security and police patrolling the streets. He said that 'there are many thugs in the streets who attack people to steal from them. Sometimes they beat them'. Police are reluctant to do anything about it and just accept it as a way of life, therefore victims do not file a report with them – they just leave it. They seldom see a policeman in the community. Another major problem in the community is the shortage of water and the disposal of waste – 'streets are stinking'.

Although people outside the slums might look down on them living in these poor conditions and 'stinking streets' (the sewage and garbage

are not collected, as the people did not pay the local government for services), they assist each other to make ends meet. David's mother had the following explanation:

> 'The community helps out if people are very hungry and families stand together. When we moved here, there were only two plots and we could buy one. We had to get used to the idea that we were no longer in California. We have to be happy here and you cannot torture your brain. We have to hope for tomorrow together.'

It is this belief in a better tomorrow and spirit of care that is part and parcel of being involved in community work and is supported by family members.

VAP and the YDF programme

Prevailing in David's personal experiences is the need for finding enough money to be independent and to assist his family financially. Joining VAP provided him with the opportunity to apply his football skills and knowledge and earn respect from his family. He describes his experiences of being a coach and how it affects his relationship with his family:

> 'I had some experience of being a soccer coach when I was still at school. Three years ago I heard about VAP and joined them. *Vijana* stands for "Youth", *Amani* for "Peace" and *Pamoja* for "Togetherness".
>
> I spend most of my weekends coaching and would sometimes be out till seven o'clock in the evenings. As a volunteer, I did not earn anything. Now with the YDF programme, I am earning and I save all I can to be able to give 1,500 shillings [13 euros / US$18] to my daughter's mother every two weeks.'

David finds recognition in his role as coach and feels that he is making a difference in the life of youths in the community.

'What is good about my life is that I feel proud that I am coaching. I get the opportunity to change the lives of my players. I give them

good advice, even with the little that I know.'

Despite the positive effects that VAP has had on David and on the community, it brought an additional financial burden. Often he will not have sufficient funds for travelling or food and has to borrow from his mother. In her opinion, being a coach should be a paying job:

> 'At times, David has to pay for his own transport to go to a place to coach or to go for a game. Then I will give him around 200 shillings [1.80 euros / US$2.40]. But his income is too small. David needs a proper job.'

Nevertheless, his family is proud of David and sees the positive sides of his involvement in VAP. 'He is one of the peer-people educating the youth about life skills and dedicates his time to others,' says his mother. His brother adds that David 'has a future. He is such a kind person, always helping people. He should just keep on going and he will be blessed'.

V.A.P OFFICE
NINE STAR
Nine Star

'I have become a survivor and can be an example for others'

Masego, Kick4Life, Lesotho

Masego lives in Motse-mocha, a small village in Lesotho. In 2006, she found out that she was HIV positive. The following year, the young woman joined the NGO Kick4Life where she was accepted and found comfort during this difficult period. Since then, Masego has been assisting as a coach and peer-educator.

Background

The year 2006 meant a turning point in Masego's life, as she was diagnosed HIV positive. Before she got the test result, she had been living a life just like many other girls in Lesotho. Her mother recalls, that Masego 'was always good at debate, music, drama and athletics. She was soft while her younger sister was cheeky'. But from one day of being diagnosed HIV positive, Masego had a negative image and many people in her community stayed away from her. Today, being HIV positive does not pose an immediate threat to her. The risk of tuberculosis is more dangerous. 'Even diabetes is worse,' says Masego's mother.

Masego talks about the occurrence that changed her life:

'It happened like this. I was at athletics in 2006. At the end of the day I could not find a taxi going to my place. I had to walk home and it was getting late. As I walked along, there was a man asking me where I was going. At first I was not scared, but then I felt threatened. He grabbed me before I could run away. He dragged me in the bushes and started to rape me. I shouted and bit him. There were two people who saw us. When he noticed that they were looking, he let me go and I ran home. Later I found out that this man had recently been released from prison – actually only two months before the incident. He had been in jail for rape. I did not tell people what happened, as I was very ashamed.'

Sometime later, Masego broke out in sores and got herself tested for HIV. She told her mother who was very upset. Even worse was the reaction and treatment from Masego's best friend: 'When her parents found out about my status, they did not allow her to come to my house anymore. Her attitude towards me changed completely.' Fortunately, Masego heard about Kick4Life:

'Their programme really saved me in that difficult time. I joined the organisation and soon became fully involved. Today, my mother is happy for me volunteering. In the beginning she could not understand why it was every weekend just Kick4Life, but now she sees how much I benefit. I decided to go for Kick4Life as they accepted me. I got the opportunity to speak publicly about my HIV status and even the newspaper wrote about me. It was useless to try to get the culprit convicted as the police were not interested.'

Community life

According to Masego, the communities around Motse-mocha are 'full of thieves and it is not safe to walk around at late hours'. Young girls have relationships with men 'to get presents and in the hope to get married'. Many children live on the streets. Children as young as 16 years start smoking *dagga*, which is very cheap. 'Sometimes people are threatened with guns,' says Masego. 'It is not a safe community.' But Masego sees a way out of this difficult environment:

'Sports organisations can help by developing sporting talent and providing opportunities. Kick4Life is well known in the area and about 10 regular participants are from our

village. I have seen participants change when they got involved in the programme. We as volunteers talk to the kids about their HIV status and they feel free to talk to us.'

Masego's mother complains that 'there is a lot of ignorance about HIV and AIDS. Many people do not understand the disease. People will hide that they are HIV positive. Some are just scared and do not want to change'.

Kick4Life and the YDF programme

Masego found acceptance, support and "a life" at Kick4Life. A scholarship programme and personal loan made it possible for her to continue her schooling:

'I got a scholarship for five years from Kick4Life that paid for my books, uniforms and school fees. The manager also lent me R300 [30 euros / US$39] to assist me in my venture. I buy sweets and then sell them in the community. I make as much as R200 [20 euros / US$26] per month, but I have to pay back small amounts monthly. I have become a survivor and can be an example for others.'

Masego's mother points out that her daughter has changed a lot since she joined Kick4Life: 'She now organises little children from the community to go for play. She has a club, the "Young Hearts", with 30 teenagers. On Saturdays and on holidays they go to old people and visit them in their homes. We collect clothes and give them to them.'

'Masego has become famous. We even had a Kick4Life event here in the community. It was called "Masego makes her move". The newspaper reported it. She stood up and spoke openly about her HIV status.'

Kick4Life has given Masego an identity – that of "spokesperson" which brings special recognition to the organisation, the family and wider community. The programme validated her existence.

She worded this embodiment as follows:

> 'I gained knowledge on how to be a skilled person – to stand on my own feet. In life you will always find difficulties. When plan A fails, there is always a plan B. The programme gives us good life skills. I want to be involved in it forever and ever because it has become a part of me and I have become a part of it.'

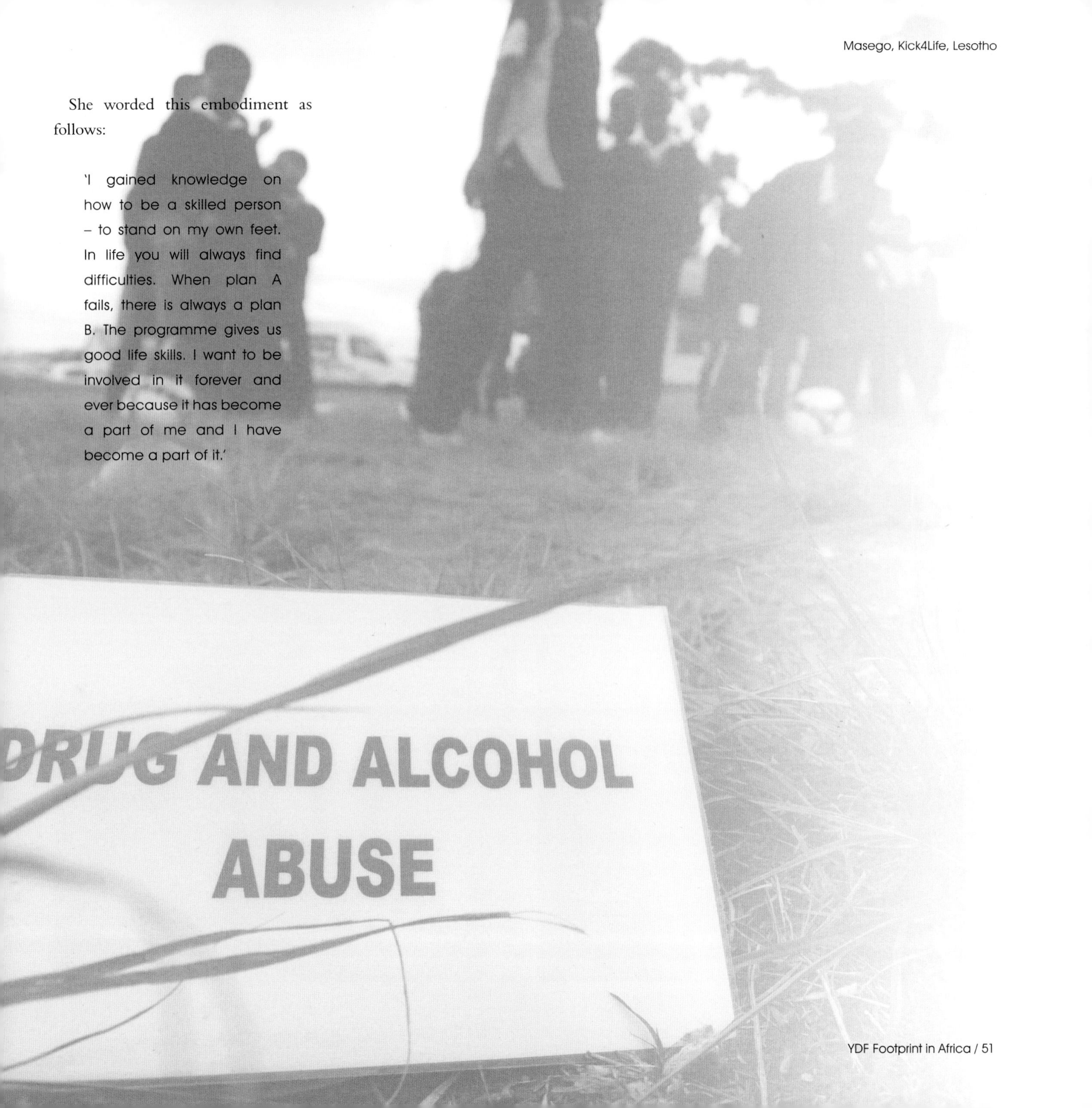

Masego, Kick4Life, Lesotho

HUMAN

IMMUNE SYSTEM

HIV

ARV

'She will never give up. The way I feel about my church, she feels about soccer'

Sarah, Namibian Football Association, Namibia

Sarah is a coach for the Namibian Football Association (NFA). She is running a football academy in the country's capital Windhoek while working full time for a telecommunications company. The 40-year-old and her cousin's family live in Khomasdal, a "coloured-dominated" area of Windhoek.

Background

Sarah lives with her cousin's family, but during weekends she visits her mother in Katatura. Sarah also has to contribute towards the expenses, as her mother only earns 150 Namibian dollars (15 euros / US$20) per month as a pension from the church where she works, and because her mother is assisting her in washing her soccer team's jerseys. 'She is a supportive and kind daughter,' says Sarah's mother.

Sarah describes her life as follows:

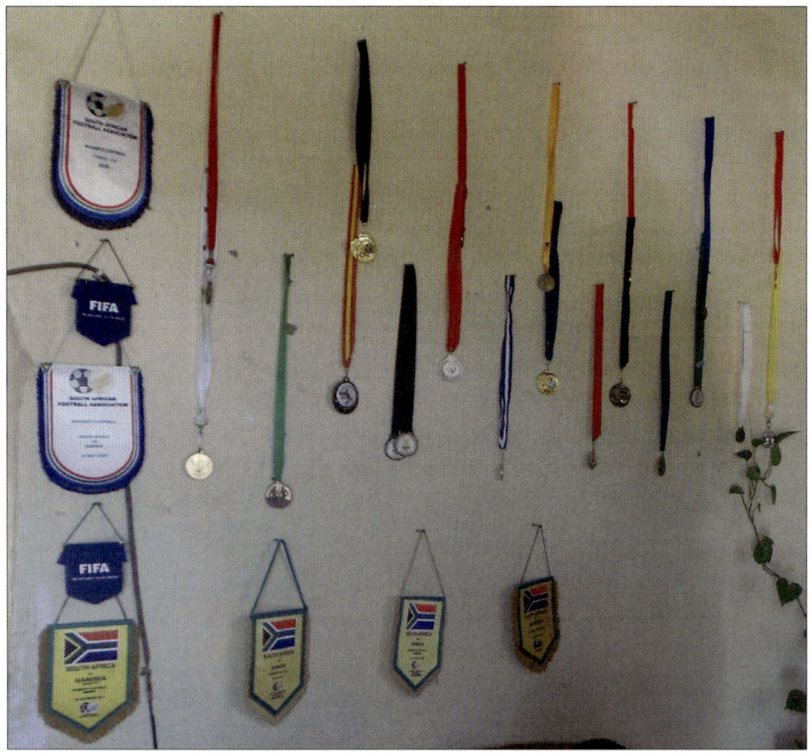

'I finished school in 1988 but could not get a bursary to study. I actually started as a domestic worker but then got a job at Telkom in 1990. I started to study the Development Management Programme at UNISA, but had to give that up. In the meantime I did several short courses at Telkom.'

Sarah would like to study journalism. But she has difficulty in finding something that she can do over a distance, parallel to working. During the 2010 World Cup and the African Cup of Nations, she was broadcasting on a Damara radio station. 'I enjoyed it very much,' she says. 'It was a three-hour programme, and people listened to me and told me afterwards that they liked it a lot. Through that, I have become the unofficial spokesperson for ladies' soccer in Namibia.'

Sarah grew up in a time when ethnic divisions were cause for local fights and she has vivid memories of those days. 'When I was in primary school we used to have lots of fights on the playground between the Namas and Hereros. The group that was the first on the playground would decide who was allowed to play.'

Sport has always been part of Sarah's life. During high school, she excelled in long distances. She didn't play soccer as 'there was no soccer for girls and we did not dare play with the boys in the streets, for we would be ridiculed and labelled as lesbians and tomboys'.

Community life

The biggest problems in the community entail unemployment, poverty and teenage pregnancies. Another issue is the difficulty for youth, even those who have completed secondary school, to find employment. Teenage pregnancies are another area of concern. It is difficult for mothers to discipline their daughters as many of them were teenage mothers themselves.

Competitive soccer, the "Academy" and the YDF programme

Sarah is very vocal about her plans for developing soccer talent through coaching, league competitions and her own academy. She explains the situation and plans in great detail:

'I am a founding member of Women's Football in Namibia. We continued during exile and our female players attracted media coverage in China and the USA. In 2006, I started the academy in Windhoek. We identified talent across the country, but we did not have facilities for education in Windhoek for all. We needed to supply accommodation. Parents cannot always afford a contribution. I have a male senior coach and I am assisting with the administration.'

Sarah is glad to have partners for her project. Her mother is involved in the project as she washes the jerseys after games. Sarah's employer, Telkom, assists sometimes. As Sarah gets recognition from the community, so the company will sponsor the medals. The YDF programme carries the league and conducts courses. The educational workshops are very valuable where they talk about not drinking or smoking, HIV and AIDS, as well as other life skills.

Although Sarah's mother is very proud of her daughter's achievements in soccer, she feels unhappy because of the financial strain it puts on their family. She has very ambivalent feelings about her daughter's dream of dedicating her life to soccer development. 'But I have started to accept that this will not change and that soccer will always be a part of Sarah's life. We were very proud of her when we heard her talking on the radio. It made me so happy. She broadcast during the World Cup and she could really say so many interesting things.'

Sarah would like her team to go and compete in other towns, but her and her family cannot afford the transport. Sarah's mother is impressed by her daughter's determination: 'She will never give up. The way I feel about my church, she feels about soccer.'

Sometimes Sarah's mother gets very sad and feels sorry for Sarah. 'The Namibian Football Association at Soccer House is always looking for her and she helps them.' However, they do not pay for her services. 'We never get any donation from them,' Sarah's mother complains. 'We even have to buy our own soccer balls. Unfortunately, Sarah brings the NFA to a certain point, but then they drop her.'

Sarah's senior coach, who joined her when he retired, is assisting full time. He also uses his vehicle to transport the players to matches and gets a small remuneration from Sarah. Despite having good qualifications and being active in community work, the lack of funding is limiting the scope of their soccer development programme and initiatives.

Sarah, Namibian Football Association, Namibia

'Sport brings people together with the universal language of play and football'

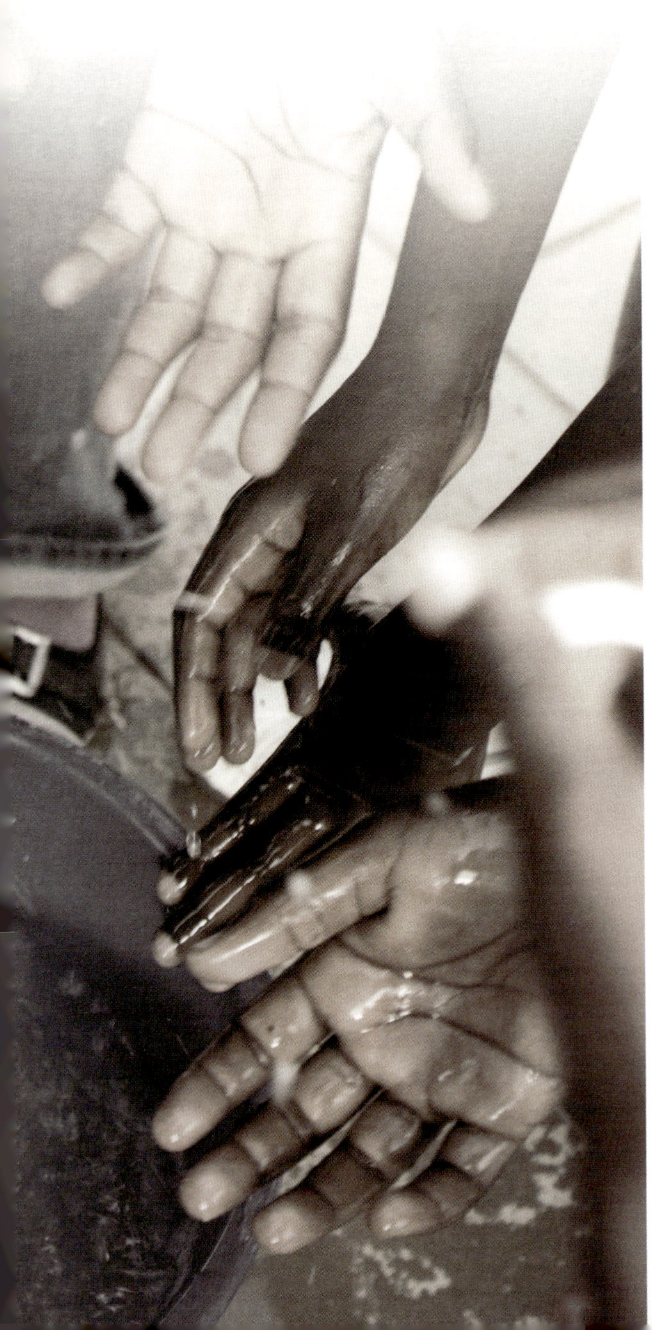

Gahigi, Chiro, Burundi

Gahigi works as a coach for Chiro, an organisation that is active in the field of youth development in Burundi and works according to its slogan: "Put away what separates and discover what unifies us". Using soccer to promote peace and reconciliation is one of its approaches. The 27-year-old man, who graduated from Burundi University, lives in the capital Bujumbura together with his younger siblings. For many years, Gahigi has been coaching a soccer team.

Background

Gahigi lives with his brother and his two sisters. Both their parents died in the early 1990s. Today, there is no one who earns an income for the household, but the parents left a house for the children, which they rent out. Gahigi went to a public university, which he paid from a scholarship. Although disagreements come up in the household, the atmosphere is good as the family members help each other out. Nevertheless, Gahigi is often concerned about the responsibilities he has to take for his siblings.

Community life

Gahigi describes community life in terms of the aftermath of the war – internally and with neighbouring nations – as their own situation that was a reflection of the Hutu–Tutsi conflict which devastated the region. He says that after the heavy conflict of the 1990s, the government tried to develop the country. 'But there are still thieves who come at night and kill people,' Gahigi explains.

The social problems facing communities mainly relate to high levels of unemployment, poverty and lack of education:

> 'There are many girls who get pregnant – they just finish primary school and live with their families. Sometimes it happens in poor families that the girl will become a prostitute after having the baby, to support the family with money.'

Gahigi said that corruption in the government is an issue as young people are expected to bribe officials to secure employment in this sector. Although he completed his law degree at Burundi University, he is still looking for employment.

In Kanyosha, where the YDF methods are implemented, the communities were very poor after the war. According to Gahigi, 'There were not enough houses or dwellings, only a few jobs and most of the livestock had been eaten. Though there was much desire for social integration after the war, deep-seated hatred between earlier warring populations remained.'

The YDF programme

Gahigi has been involved as a coach for several years and is responsible for a team of youths and young adults between the ages of 14 and 25 years. He is very excited about the YDF training and is looking forward to Chiro's upcoming training for the "Peace Games".

He was also involved in organising the "Peace Caravan", a series of events touring across East Africa bringing together soccer teams from Burundi, Uganda, Kenya, Rwanda, Tanzania and South Africa in 2009. The events aimed at promoting the approach of peace through sports. Gahigi is now 'dedicated to become a trainer in my community and implement the YDF methodology'. According to him, the "Football for Peace" approach has a vital role to play in addressing conflict, equality and reconciliation in the communities:

> 'Before the implementation of "Football for Peace", many young men were ashamed of their role in the war – either as victim or as perpetrator. There are high levels of guilt among people relating to their roles or the absence of

lending help to those affected by the war. The main focus is on the rebuilding of value systems by working with parishes across the country.'

Chiro's football programme reaches nine communities and is implemented by 70 coaches. Its main objective is to offer education and develop social values with the emphasis on brotherhood. The impact of the training is the building of the coaches' capacity to deliver "Football for Peace" activities. Through the training, the workshop participants gain knowledge and confidence to use the YDF methods in their peace work. Gahigi explains that GIZ supports the Youth Centre in Kanyosha by making funding available for some local initiatives. He adds: 'For our organisation which got the idea of "Football for Peace" from South Africa [YDF] and Kenya [MYSA], it was a great tool to build on the initiatives.'

The popularity of "Football for Peace" as a movement is evident, as it became the flavour of peace programmes with about 400 ex-soldiers among the 12,000 beneficiaries since the year 2000. This programme was also integrated into existing initiatives to address the destruction of war on human lives and to warn future generations about war and conflict.

There is now representation from all the groups in the government, but it does not answer local needs about reconciliation. Issues of discrimination and gender inequality are also being addressed. Previously, women were not supposed to take part in public life and had no place in football. The focus of delivery is to train male and female coaches and to focus on fair play, which is most relevant for war-torn countries.

For Gahigi, "Football for Peace" also includes the teaching of life skills and of including girls in teams to try and change the gender disparities and perceptions of how men and boys see their female counterparts. Other issues on the agenda include care of the environment and the fight against HIV and AIDS.

Gahigi said that 'young coaches should also be equipped to be creative and not be dependent on what they can learn from others; they need to be able to think for themselves and answer the needs of their participants'. Sometimes, he says, "Football for Peace" is not enough:

'Some feelings are so deep-seated that it is not so easy to address them through football. We have had success by putting together people from opposing groups into one team and when the hate and violence take over, the coach will intervene. It is difficult to forgive somebody who wrecked your life.'

FEAR
R282 G
GH-STOP

'Now I know better how to approach life'

Mike, EduSport, Zambia

Mike is a peer-educator and coach who lives in Livingstone, Zambia. The 22-year-old is affiliated with the NGO EduSport, which seeks to foster community education, development and empowerment by using sport. The concept is based on the belief that sport with a positive orientation is a powerful vehicle for change. Mike, who lives with a friend, tries to raise the awareness for health and hygiene through football.

Background

Mike currently lives in the Libuyu compound section of Livingstone with his friend. Their one-roomed "house" is on the premises of his sister's dwelling. They have a double bed, four chairs and a table. Running water is available outside, but there is no electricity. Mike's sister makes a living by selling fresh produce which she buys weekly from farmers.

Mike's story is full of hardship and human kindness as he has to fend for himself, being assisted by his only sister, friends and people who give him odd jobs to earn some income. His recollections mostly centre around fending for himself from a very young age:

'We grew up in a rural community. I went to a local school that they selected for me but it was getting expensive. Then I did not get very good marks and went to another local basic school. I did not have a sponsor.'

Mike's father was accused of fraud by the company that he worked for and went to jail. After six months he was released but stayed unemployed.

'My sister and friends paid for my school for the first term in Grade 10. In exchange I did little jobs for them. Finally, I came to stay with my sister in the compound to find jobs to pay for my school. I made small businesses and sold things for others.'

One of the main challenges of 'people without money' is that they cannot afford school fees. This perpetuates the cycle of poverty as it restricts them to menial jobs and trying to make ends meet by engaging in the informal market. For Mike, the account of his life is mostly about being able to survive and to complete his schooling, with the dream of being able to study further that might in the end lead to stable employment.

'It was not possible for me to go and study because we have a strict system. I got 17 points which was very high. In Division I you can have between 6 and 20 points. In Division II you can have between 20 and 25 points.

My results are still being kept by the school and I cannot apply for a job. I first have to pay 645,000 kwacha [95 euros / US$125] to get them to release my marks. My cousin completed his schooling in 2001 and only could pay to get his results in 2011. When I was in Grade 10, food was very expensive and sometimes we would go without food for a whole week. I might get some lunch from my friends. My sister would then just stay without food. The teachers also assisted as much as they could. They taught me about life and encouraged me to concentrate on my education.'

Mike's sister found a way out. Despite her lack of education, she married a man who is employed. She is very kind and looks after Mike and their mother as much as she can.

Mike played a lot of sport during his primary school years. He said that he loved sport as he was 'born with it'. In basic school, he played football, basketball, badminton and volleyball. Playing sport made him feel 'free as a child'.

As he was clever, mostly being second in his class, he had lots of friends who wanted him to help them:

> 'Friends were very important. Without friends we would not have survived. However, when I was in primary school I played a lot. I played with naughty boys and did not have a guide. When I came to high school, I made different friends. Now, my friends are in the YDF programme. Desirée, who is also a peer-educator, told us about EduSport and this programme. We were eager to join as you still have time to do some jobs for money.
>
> As coaches, we are close. You get different perspectives from different people. If you can be in line with what they are thinking, you can select what to talk to them about.'

Community life

The daily struggles of life in poverty often consume young people who can from time to time escape by playing sport, or find great satisfaction in small happenings and jobs where they can earn something, enjoy being with friends or family members, and also contribute to relieving their own hunger and sharing the little they get with others.

> 'Most people are too ignorant – they drink a lot and there are many young girls that are pregnant. I know of more than 20 girls who do prostitution to survive. Their parents know but are just glad to get the money.'

The YDF programme

Mike views himself as a coach for football and a facilitator for "Health and Hygiene". YDF has been an educational venture for Mike:

> 'Kids get information about health and hygiene. With boys and girls in the same team, they learn a lot and grow up. The sanitation in my community is especially bad. GIZ built toilets and teaches people how to maintain them. In Libuyu, an older area where the poorer people live, they need knowledge.'

Mike's friend provides detail on their earnings as volunteers for YDF and how important this income and a bit of food are for their survival:

> 'I am now a volunteer and earn no money except when I facilitate a workshop. I presented six workshops of two to three days and earned 30,000 kwacha [4 euros / US$6] per day. I also get a travelling allowance when I do workshops. I present a workshop maybe every three months. At the matches that we play every weekend, I get food. The teams play matches for the league every weekend – then I get a transport fee of 10,000 kwacha [1.50 euros, US$2] and a meal.'

Both of them would like to remain involved in the programme, but are driven to find employment to merely survive. Mike's sister's perception of his involvement with YDF is that he is 'doing a good job', but that this is not meaningful enough given the poverty and needs they face daily.

Lerato, Altus Sport, South Africa

Lerato is a peer-leader and coach for Altus Sport, an NGO that uses sport for youth development. The 19-year-old lives with her parents, her twin sister Patience and her three brothers in Mamelodi township near Pretoria. She has been passionate about soccer since her childhood. Today, she plays in a women's team and coaches youth teams.

Background

Lerato spent the first years of her life in a rural area in the province of Limpopo, where there was neither electricity nor running water. However, Lerato's family moved to Mamelodi to find better employment. The family is very close and the twins consider each other as 'my best friend'. Lerato tells about their childhood years:

> 'When we got to Mamelodi, life was happy. We played a lot of soccer with our brothers and with my father. Most of our time, we would spend with each other and with some friends in our street. At school one year, I was captain and my sister was vice-captain. The next year, we changed places – she was captain and I was vice-captain.'

'I love sitting down with the kids and talking to them. I love bringing smiles to their faces'

Patience said that the boys did not like them playing soccer, but because they played a lot with their brothers and were skilled players, they were accepted. Some boys would say: 'Wow! That's why you want to play, because you're good!' In the beginning, their father did not like them playing soccer. However, he accepted this and the whole family would support them. They have always been active in playing soccer together and socialised with boys as "soccer players". Lerato explains:

> 'We mostly played against boys which made us very good players. In the beginning we lost and could not keep up with the boys, then we founded the Simuye Ladies. There were only four girls in the team and the rest of the team were boys. The coach was firm and did not want to change the team, but wanted to develop girls and boys. The boys did not like the name, but they could not change it.'

Lerato said she and her sister became inseparable and even went as partners to the Matric Ball – 'People thought it strange but we did not care.' They would do most things together and would call each other *Mugani*, which means "friend" in Zulu.

They also became more active in the community team, named *Dlala Ntombazana*. The manager of the NGO that runs the team became Lerato's mentor. It was this manager who introduced Lerato and Patience to members of Altus Sport, and who would offer them the position of peer-leader.

Community life

For Patience, Mamelodi is a good place. People help each other, there are nice parks, and 'good people around, children support with everything you do'. There is a vibrant community life and people from Pretoria will travel to Mamelodi to "party".

Patience says she does not like *tsotsis* (bad people) who come to break in, steal, take water pipes and run away. 'Some boys smoke, steal taps, leave the water running and then sell the taps to buy more drugs. The boys who smoke in the park teach the younger kids the same thing – they are negative role models.'

The YDF programme

Since they joined Altus Sport in 2007, Lerato and Patience first implemented the programme in a nearby park in Mamelodi, but then Altus Sport decided that they should offer a more focused programme in the schools. In 2007, Lerato was selected to go to a streetsoccer tournament in South America:

'I got training and I was chosen to go to Paraguay. I was not necessarily the best player, but they took me for my leadership qualities and personality. It was a sad moment for my sister, as she could not come along. I did not tell anybody and only told my mother two days later after an evening meal.

During the week in South America I got to talk a lot about my community and my country. It changed me and I realised that I can be somebody in sport. I was not shy anymore and could speak in front of people. Even now, I can just start talking to somebody that I do not even know.'

In 2008, Patience got the opportunity to go to Chile as a representative of Altus Sport. She confirmed what her sister said and refers to the 'disappointment and pain to be left behind', yet to be later chosen to go to Chile 'eased the pain' and brought the sisters together again. Lerato talks about her experience as a youth leader:

'I love sitting down with the kids and talking to them. I love bringing smiles to their faces and then asking them what is wrong with them if I see that they are sad. Some do not have easy lives. I would also tell them how people made fun of me, calling me a lesbian just because I was playing soccer.'

Lerato said that being called a lesbian has become part and parcel of playing soccer. However, when they play against other teams, the "real" lesbians will try and 'hook up with us by asking our phone numbers'.

Lerato is full of praise for the opportunities that she got through Altus Sport which aided her personal growth and provided her with direction in life:

'My work with Altus Sport afforded me lots of opportunities. I have been given so many experiences to develop personally. Last year I was sent to Germany, where I attended a seminar on how to follow one's own path of development. We got career guidance and were told how to go about it to reach our goals.

GIZ gave us opportunities to travel and to attend short courses. We were trained how to deal with journalists and many of them interviewed me during the World Cup. I attended various panels and did public speaking.'

The twins both consider studying the National Diploma in Sport Management at Tshwane University of Technology, as they were not successful in their application to the University of Pretoria last year. If they get accepted, they will inevitably leave their "jobs" at Altus Sport – something that parents will regret as 'the kids are happy around them and it is not easy to find people to trust to take over'.

'I have learnt patience. It is kind of cute to work with children. They often ask for me and it makes me very proud'

Candice, Altus Sport, South Africa

Candice is a schoolgirl who lives in Mamelodi, a township near South Africa's capital Pretoria. For the last two years, the 14-year-old has been attending life skills sessions organised by Altus Sport at her school. Recently, she has also been assisting in the NGO's "Life's a Ball" programme. Candice lives with her mother, her stepbrother and her uncle's family.

Background

Candice was born in Boksburg and grew up with her mother and father. When she was five years old, her father committed suicide but nobody knew why. Currently the household consists of Candice, her mother, her uncle, his two children and her stepbrother.

Candice is very fond of her stepbrother. Her mother's boyfriend does not want to get married, but provides some income as he is a 'rich businessman'. As soon as her uncle marries her aunt, they will move out. Candice will miss her aunt, as she can talk freely with her. She explains: 'I cannot talk like that to my mother as she is always shouting at me. My relationship with my mother is not so good. She doesn't understand my life.'

Candice's mother is very sad that she and her daughter are not close. However,

she says that Candice is her 'left hand as she takes care of the house money, cares for her brother, cleans and cooks'. She is proud of Candice because she is doing well at school by scoring high marks in different subjects.

Lives intertwine, yet follow paths of enclosed personal meaning and experiences where each household member will have a different understanding and perception of shared happenings. Candice's mother would like to live on her own with her two children, but finds it hard to leave her parents, especially since her brother is moving into his own home.

Candice played netball in primary school and was a centre. Her mother also said that she took ballet during the year when she was attending a crèche in Boksburg, but since they moved to Mamelodi West, she has not taken part in any organised sport.

Community life

Candice's mother said that the 'community is cool' as there is relatively little crime and there is 'much socialising going on – we sit and gather as friends too much'.

A negative aspect in the community is not having a community hall, since the current one was vandalised and no sport can take place. Also, there are only soccer teams for men and nothing for women. There are people 'who do crime, steal, break into houses, rape and rob. Many teenagers are involved in drugs and alcohol abuse'.

The community is not clean and there is much paper and filth lying around in the streets; especially around the taverns, people will litter extensively and just throw bottles or beer cans into the street.

All family members who were interviewed refrained from talking about their own tavern as it being a possible source of concern for other community members. They would refer to the 'big tavern across the road that is noisy and where many people go to drink', whereas only Candice and her nephew said that they would ask their uncle from time to time to turn the volume of the music down in their own pub so that they could study or sleep.

The YDF programme

Candice's mother said she had limited knowledge of the YDF programme and mostly had heard only about Altus Sport as she had to sign their forms to give permission for her daughter to be involved. She conveys the information she has about her daughter's involvement:

> 'At Altus Sport they are playing soccer and other sports; they do gymnastics, go to primary schools to help other

kids read and write. Candice has been going to the programme since Grade 7. The first day after having joined, she came home and said her muscles were stiff, that they hurt. I asked her why and she said she was gyming and that tomorrow she would be late because she was going to another school to help other kids read and write. I had to fill in forms to release her to participate. I think it is good, nice and I love it. Altus can help keep the kids off the streets – kids are clipped by cars because they are on the streets. Altus will keep them occupied, away from illegal things, like stealing. They help take the teens' mind off those things; they are tired after school.'

Candice is enthusiastic about her involvement in teaching other kids to improve their reading and writing. She said she also likes the sports, and she particularly enjoys the life skills sessions where 'you can take leadership and set an example, and also where you can express yourself and listen to others' opinions.

'Being involved with this programme changed me. It gave me confidence and I can stand up in front of a bunch of people. It was not like this before. I also know you must set a limit with boys – you can be friendly, but must learn to keep them out.'

Candice's mother said that she did not observe any change, except that her daughter is busy in a good way. She likes to help others and when she returns from the schools where she teaches the Altus Sport programme "Reading & Writing", she is happy and diligently does her homework before going to sleep relatively early. As a mother she is thankful for such a programme that can keep her daughter busy and 'out of trouble' and 'it refreshes her mind'.

For Candice, the involvement in the "Life's a Ball" and "Reading & Writing" programmes holds the following personal benefits:

'I mainly learnt how to play soccer and other sport skills. I can now pass a ball in rugby and also know the rules of cricket. I also watch these sports with my uncle and we enjoy it together. We support the South African teams.

By being involved with "Reading & Writing", I have learnt patience. It is kind of cute to work with children. They often ask for me and it makes me very proud.'

STARTS HERE

STARTS HERE
86

'Soccer is Tumi's life – even when she was little ... If you take soccer out of Tumi, you would kill her'

Tumi, Dona's Mates, South Africa

Tumi is a high school student from Orange Farm, a township south of Johannesburg. The 19-year-old has been playing for Dona's Mates' football team for several years. Tumi has always felt more like a boy than a girl but had to struggle with prejudices and offence in her personal environment. Today, she is openly lesbian and feels comfortable talking about it.

Background

Tumi lives with her mother and two brothers, originally being a relatively large family, but four children have passed away. The family moved to Orange Farm in 1999, but for several years Tumi stayed with her mother in Randpark Ridge, where she worked as a domestic worker. Her father left them long ago. Nobody works in the house. Her grandmother sometimes came to stay and look after them, but when her mother retired, she came home to live with them. Central to Tumi's life is her love for soccer and that she is openly lesbian:

'Many people said that girls who play soccer act like boys, and that they are lesbian. They said that these lesbians will rope everybody in and the fact that the girls like trousers and not skirts, makes them be like boys. They have to carry the stigma.'

Her mother says that it was very difficult for her in the beginning to 'cope with Tumi being like a boy', and tells about her own journey of accepting her daughter the way she is:

'People would call her names behind her back and to her face. But now, she has friends that look like her so it is more acceptable. When Tumi was in primary school she didn't want to wear the skirts or dresses the girls were supposed to wear. She would throw her panties in the toilet. We would fight every day. Tumi would prefer boys' shorts, not a dress. When she was asked if she was a boy or a girl, she would say she was a boy.

Tumi was not comfortable playing with girls; she didn't have any dolls. One of her sisters bought her a dinner set to play with when she was four years old. Instead of playing with it, she dug a hole and buried the dinner set. This was really difficult for our family.'

At school, Tumi's behaviour caused problems for her and for her mother, as one teacher placed the blame on her mother. 'I had a discussion with a teacher who said Tumi needed to change and that I am not doing my job as a mother.' Another teacher heard about the discussion and referred Tumi's mother to a social worker, who said that the family needed to accept Tumi the way she was.

Initially the family thought her being lesbian was because she was playing soccer with boys. She would socialise with them and they would come and visit to talk about soccer. In high school she found friends like her: girls who played soccer and dressed like boys. Some of these girls still deny the reality of being lesbian but Tumi was open from day one. Other girls were "in hiding", but when they saw Tumi, they became friends and were more open.

Tumi's eldest sister is less 'impressed' with Tumi's lesbian status. She says she loves her sister and has plans to give her a nice 21st birthday, but she is not happy that Tumi wants to invite her 'girl friends'. For her, girls are girls and boys are boys. She does not want to accept the fact that Tumi is like a boy.

Community life

Problems in Orange Farm, mentioned by Tumi's mother and brother, include poor infrastructure, unemployment and crime. They are waiting for the government to help. 'The youth sit on the street doing nothing, but there are so many volunteer opportunities that could get them somewhere,' they say.

In Orange Farm, young girls often get pregnant at the age of fourteen. Tumi's mother contributes this to unemployment. She says girls sell their bodies to drunkards and to taxi drivers who have cash to pay them or give them gifts. Many young people brag about their sexual exploits. On this topic, a coach from Dona's Mates explains how girls and boys would 'take competition' on their conquests:

> 'Girls are very into boys; they might even say that they have the whole cabinet to give them presents. They might say: "I have the minister of hair styles, the minister of transport and the minister of cell phones." In turn, boys like to brag that they are dating girls; they might say: "I have a girl for breakfast, for lunch, for a snack and for dinner."'

Tumi never openly admitted that she was a lesbian, but just said that she is put off by the behaviour of girls who always talk about boys and by girls getting pregnant. She doesn't like how girls behave. Girls gossip too much for her liking.

Dona's Mates and the YDF programme

Tumi's mother thinks her daughter has a bright future in terms of sport. 'If she puts her mind to soccer she will get good results. Our family helps her with her education. She is a slow learner. But she can't just do soccer because when she gets old she can't kick a ball anymore.' Yet, at the same time, she says that Tumi should get a job working with her hands. 'It's not easy here. Kids are used as soccer players but when they are no longer good, the club just leaves them.' Soccer can lead to disappointment. The fact that Tumi's mother has decided herself that her daughter should not focus on soccer is related to an incident where Tumi sustained a major injury:

> 'In 2009 Tumi broke her leg playing soccer with boys. We had to take her to hospital and she had an operation. They

> put a pin in. Sometimes it still hurts, especially when it's cold.'

Tumi's brother thinks she was 'kicked on purpose by a boy who did not like her playing so well and had something against her being a lesbian'. It seems to be a 'career-ending injury, for she cannot run properly and sometimes limps on the field'. The coach says that he had instructed the other players not to attack her, but to make it possible for her to play with them.

Tumi's mother is very grateful for the work that Dona's Mates does in her community:

> 'Dona's Mates is good. They expose children to many things, Dona takes the children out. The organisation is doing a good job taking people from the streets and doing something else with them.
>
> Soccer is Tumi's life – even when she was little, she would go behind our backs. If we refused, she would throw her bag out the window and sneak out to the field and the boys would then see her at the field and ask how she got there, but they wouldn't tell her to go home. If you take soccer out of Tumi, you would kill her.'

ATES

adidas

'She gives children healthy advice. She also gives me advice, even though I am older. She is wiser than me'

Thoko, Umzingisi Foundation, South Africa

Thoko is a peer-educator and coach for the Umzingisi Foundation, an NGO which is operating in and around Port Elizabeth. The 23-year-old's passion and talent for sport made her join the organization in 2009. Thoko, her daughter and her sister live in the township of KwaZakhele on the outskirts of Port Elizabeth.

Background

Thoko is a 23-year-old female peer-educator for Umzingisi. She grew up living with her grandmother and her sisters. Her grandmother died in 2007, her father shot himself when Thoko was 14 years old, her mother passed away in February 2010 and one other sister died after an operation in February 2011.

When Thoko grew up, she had a relatively easy life as her mother worked hard and gave her children what they needed. Thoko's friend, who lives in the same neighbourhood, says that Thoko was a very active and responsible child. She remembers the time when Thoko's father committed suicide:

'Thoko's father passed away when she was in high school and this affected her very much. All of us noticed that she changed. She became quiet and withdrawn. They had depended on the father as a breadwinner and Thoko was very close to him.'

In 2006, Thoko's sister was diagnosed HIV positive. Soon, she started to take antiretroviral medicine. Today, she regularly goes to a support group where the people motivate each other, then they go out and motivate others. Thoko's sister says that 'people do not share that they are HIV positive because the community stigmatises them'. She talks about her relationship with her child's father and about being HIV positive. 'My boyfriend is HIV negative, but he does not care that I am positive. We use condoms when we have sex. I can have more children, but I only want to have two – a boy and a girl.'

Thoko wants to use her sister's status as an example to others and to become an AIDS counsellor:

'My way of teaching children is to introduce myself and tell them that my sister is HIV positive. We call it "The coach's Story". At first, they are surprised but then they realise that I can teach them better because I am in this position – it encourages them to share their problems. The kids often tell me about rapes. But they just want to share; they don't want anything to be done about it.'

Community life

When Thoko's sister and friend talk about "community life", they highlight the positive aspect of relying and depending on one another to create close-knitted "units of care" against the backdrop of many types of anti-social and destructive behaviours. Thoko's sister says that KwaZakhele is 'an average community to live in – there are good people, criminals and every kind of people in it'. Many people are affected by HIV. It is poverty that leads to crime and alcohol abuse. Boys and girls from 14 years go to shebeens. Thoko's sister recalls 'break-ins – mostly at night, but also in the day when there is no one at home'.

Thoko's friend points out that 'half of the people in the community will go to church and the other half likes to drink. There are too many shebeens and taverns. People go to drink on Sundays too – but then the shebeens close at 10 o'clock at night'.

The YDF programme

Thoko was good at athletics during her school years, and also sang

in the school choir. After school she played handball and did Thai boxing. Her love for sport and her personal circumstances influenced her decision to join Grassroot Soccer as a volunteer and to become involved in coaching with Umzingisi.

Thoko started to work for Grassroot Soccer in 2008. A year later, she joined Umzingisi as they were able to pay a stipend. She worked as a volunteer and then applied for the YDF coaching job. 'I love working with kids. I am thankful for my current job because I can earn money,' Thoko explains.

In addition to being a peer-educator for life skills, Thoko has a special responsibility. She has to 'oversee other coaches to make sure that they keep up with the programme'. She also implements "Training of Trainers" for other organisations in order to spread the YDF approach. She really likes the programme as 'you are your own person. They do not push you – they let you do your job without pressure'.

The greatest benefit for her is that she has acquired special knowledge about HIV and AIDS, as well as a new methodology. The main skills mentioned by her relate to computer literacy and how to go about planning and implementing events. The fact that they can voice their needs and concerns is evidence of the democratic leadership of Umzingisi. Peer-educators, however, mostly identify with the participants and see the organisation as a main resource for providing according to the identified needs – these of the peer-educators and of the participants.

Thoko's sister says that by being part of Umzingisi, 'Thoko acts more responsibly by rather staying at home than going out to party with her friends, and looking after her child. She also shares information and lessons from the programme with me and visitors'.

Thoko's friend has a similar perspective on the personal attributes that Thoko brings to the programme:

'Thoko is good in motivating children. She gives them healthy advice. She also gives me advice, even though I am older. I will fight with my brother and then there will be a bad relationship. She will give me advice on how to handle him and after not talking to each other for a week, we will make up. She is wiser than me.'

GRASSROOTS
PROGRAMME

'I was not used to socialising with people from other cultures. Sport helped me fit in'

Thandiwe, YDF Western Cape Network, South Africa

Thandiwe is a coach and life skills coordinator within the YDF Western Cape Network. Sport has always played an extraordinary role in the life of the 23-year-old. He lives in the township of Langa in Cape Town, but most of the time Thandiwe works in Khayalitsha, one of the roughest townships of South Africa.

Background

Thandiwe stays with his parents, his little brother and little sister in Langa. The family lives in a relatively big house and owns a tavern. Thandiwe's mother had a stroke recently and is not working. Her husband is a trained manager and has been working for *Shosholosa Mail* for four years, where he earns a 'good salary'.

Thandiwe's mother says that her son was 'lazy' and not really very helpful in the pub. She would prefer for him to leave home and settle elsewhere, as they have to employ more people to assist with the business. But Thandiwe says that his mother 'has been affected by 'bad medicine'. We have a good income from

our tavern. People are jealous, especially those who are the opposition down the street.'

Thandiwe received a Diploma in Sport Management from the Cape Peninsula University of Technology (Technikon). He finished his studies in 2009 and currently earns R3,000 (295 euros / US$390) per month from a YDF stipend. His story is very much linked with him playing sport at school and for a professional club. He tells about his experiences against the background of cultural divisions and provides a "look into the life of a soccer coach":

> 'I went to Observatory Primary School and it was mixed. People spoke English. Before, I only spoke Xhosa, so it was difficult to adapt to English. I was also not used to socialising with people from other cultures, but sport helped me fit in. In the beginning I didn't get any recognition and could not pop into a team. I played during lunch time and then the senior players saw me. I was so shy, but when they asked me to play with them, I joined in.
>
> The next morning I was pushed around in the lines, but I was not provoked to fight. I just laughed. I was big, so they could not slap me, and I started to play with them in the afternoons. Later, I went to Woolmer High in Woodstock for Grades 8 and 9, and to Maitland High for grades 10, 11 and 12. Woolmer High only had two classes and it was just too small. They only had a small field and we played street soccer. Maitland High was in a Coloured area and if you were not protected by groups, you would be hurt.
>
> I then played in the Hellenic team for the Under-17 team. They went bankrupt just when I started to play for their Under-19 side. The club was bought and I went for trials. I did not make it and then concentrated on my academics.'

Today, Thandiwe is a member of the Youth Forum of the YDF Western Cape Network. He is also a life skills coordinator for Amandla, an NGO in Khayalitsha:

> 'We are 12 young leaders and deliver the "Heads Up" programme on Mondays for 32 teams. We play 5-a-side football with eight or six players per team. We play a log for the Heads Up league. In the same game we also keep a Fair Play log. We also referee and get all the teams together and explain about fair play and how the scores work. With the Fair Play log the players behave by showing respect and shake hands after the game.'

Community life

Thandiwe and his family are positive about the support of friends, however experience some jealousy of others. Thandiwe's mother is also 'happy when neighbours assist with washing and cooking which I cannot do because of my illness'.

It is mainly Thandiwe who talks about the challenges facing youth in the community. Boys are exposed to drugs, gangs and violence, whereas girls often become 'young mothers before their time'.

The YDF programme

After obtaining a national diploma in Sport Management from Cape Peninsula University of Technology in 2009, Thandiwe also worked for the Department of Sport and Recreation of the Western Cape province for three months in 2010. He was a Facility Coordinator and intern in Kuils River, but his contract was not renewed and he had to look for employment elsewhere. He tells about his becoming involved with the YDF programme:

> 'In February I saw the advertisement for YDF and went for training for four days. As a professional player I am used to setting goals. If there are barriers, I am used to looking for alternatives. I set new goals – now I would like to become Chief Director and want to work for the government. Before that, I would like to have my own programme in Langa. I would like to give back to the community as I am fully aware of how to mobilise the youth and how to promote the youth.'

He is very positive about the example he sets and the role he is playing in improving the lives of young people. He explains:

'Currently I am a role model to the guys. I am like a father to 110 kids. I have to look after them and have to take their minds away from their home circumstances. Most are abused children. Many of them do not want to go home. They need to learn to respect themselves first. I became more exposed to life situations. Now I know better how to approach life.'

Thandiwe, YDF Western Cape, South Africa

'We can shine in the darkness if we help a child. This is my calling'

Lesego, SEDYEL, Botswana

Lesego is a university student from Ramotswa, Botswana, who has been working for the NGO SEDYEL as a coach and public relations intern. She joined the organisation in 2008 as a peer-educator and set up a drama group that performed several plays. Today, she still coaches a girls' team.

Background

Lesego is attending the University of Creative Technology Botswana. She says that many people find it hard to understand why she goes to university when her family is actually poor and suffering. The reason behind her academic efforts is that one day she found a journalism programme on the internet. 'I set my heart on this and just wanted to go.'

Currently Lesego is doing an internship at SEDYEL. 'I am writing their newsletter and creating a website for them. I am also assisting with the establishment of a photo gallery which will help their PR Department,' she explains. Lesego likes the spotlight and dreams about being on TV.

Lesego's mother is 41 years old and works as an accountant in Gaborone. Her father and brother are both mechanics and also work in the country's capital. Lesego doesn't have contact with her father as he left the family and remarried.

Lesego's mother provides further information about the household and sketches her own background as follows:

> 'At the moment the household consists of me, my three brothers, my four children and my daughter's son. Luckily I don't have to pay any more school fees for Lesego – the government is paying. She finished a three-and-a-half-year course at the University in Broadcasting and Journalism in 2010. When her results come out in August, she will graduate and apply for a job.'

School years and community life

During her school years Lesego excelled in sport and drama, passions that she would retain after leaving school and in deciding on a career as journalist. 'In primary school,' Lesego explains, 'I played netball, threw javelin and took part in drama.' The drama group especially became a big part of Lesego's life:

> 'I played for a community group which was very exciting. We had to act out real-life situations and had to imagine that we were in somebody else's shoes. The drama group is still there and sometimes they will have a show during a *kgotla*, a meeting of elders in the community. Passion killings that affected our community became a famous theme for them to perform in the community and at schools. Nowadays most plays are about HIV and AIDS.
>
> We had a show about the girl-child – "I am the future". It is not good that people still have the stereotype about women coming second to men. Girls can do better. Ladies can also be heroes and should be taken seriously. They can do the same work as men.'

Asked about community life, Lesego explains that 'in Gaborone, it is everyone for his own business. They do not help each other. Ramotswa, in contrast, is a quiet place. The most fun we have is attending football matches and going to church on Sunday'.

SEDYEL and career development

Lesego took her first steps at SEDYEL in 2008. 'That year, I became involved as a peer-educator and put a drama together with the title of "Like father like son". Another drama was called "*Sejosennye*", which means "little food but you can do better"' I found a comfort zone with them as they accepted me and gave me the opportunity to assist. I love being a mentor in my spare time. It makes me feel valuable. I can also

make up my own girls' team. We can play football and talk about life.'

When she grew up, Lesego had a very low self-esteem and acted very shy. 'I did not want to be exposed to other people,' she recalls. Yet, having met with overseas people through SEDYEL and seeing that they have programmes in Africa lifted Lesego's confidence. 'I saw on the internet that they have scholarships for African women. This showed me that being from Africa does not mean disqualification.'

Lesego likes the fact that football can open up many opportunities. 'Sport brings people together with the universal language of play and football.' Lesego is proud of her SEDYEL team, which is young, but the players are very open with her. She points out that 'they do not bottle up their feelings and sometimes they even cry. I also cry about their and my things – it is a good remedy when you are sad. They can talk to me and I understand'. Then she talks about the challenges of her players:

> 'The Under-13 girls grow up fast. They have been exposed to sex and need information about sex and about HIV and AIDS. I give them knowledge and they ask me what they want to know. They have lots of pressure from young boys who pretend to love them. Then they fall for that as they really want to be loved. Many of them have little self-confidence. Some are abused at home, but at SEDYEL we give them respect. They just want to belong and we embrace them.

Lesego comes across many girls who are very shy and some of the small girls get beaten by the boys if they do not want to 'give them love'. The girls come and talk to Lesego about this bullying and violence. Though she realises that her assistance is limited and that she needs further training, she feels proud that she can already give advice to the girls. Her own participation and assistance in SEDYEL are a source of dreams for Lesego.

> 'The world is not so big anymore for a poor girl from Ramotswa. I can contact people from other countries via the internet; we can have an active website and the possibility of studying overseas is not so impossible. I would

like to become a master trainer. Currently I am a peer-educator and still have to do level one and two, before I can be a master trainer. SEDYEL's programme "Kicking Aids Out" gives you this opportunity and recognition. We can shine in the darkness if we help a child. This is my calling.'

Lesego, SEDYEL, Botswana

TIMBITS
15

Carlsberg
DIADORA

Conclusion

Kelly Mkhonto, Director: Community Sport, SRSA, and Gerald Guskowski, YDF Project Manager, GIZ

Looking back at the road travelled with the Youth Development through Football (YDF) project brings many memorable moments to mind. The 2006 FIFA World Cup™ in Germany had just come to an end and the spotlight was about to swing onto South Africa, when the German and South African governments decided to enter into a partnership to make the 2010 FIFA World Cup™ in South Africa an unforgettable success.

South Africa was about to host the first FIFA World Cup ever to be staged on African soil. This filled not only South Africans, but the whole continent, with great pride. At the same time, expectations started growing every day that the World Cup drew closer.

To rise to the huge expectations, the South African–German partnership designed a broad array of joint programmes, ranging from infrastructural development to holistic event-management systems and construction of soccer pitches in disadvantaged communities. Additionally, YDF was envisaged to provide a social legacy to the partnership and the 2010 FIFA World Cup™ by targeting young girls and boys from across South Africa, as well as from nine other African countries as far afield as Rwanda and Ghana.

From the day the Department of Sport and Recreation South Africa (SRSA) and the Deutsche Gesellschaft für Internationale Zusammenarbeit (GIZ) GmbH launched YDF in 2007, an impressive number of things have happened, and while recounting all would most probably occupy a book on its own, a few moments deserve mentioning. In 2009 the "Peace Caravan" tour in the Great Lakes Region and the "Let's Confed, Africa!" programme in South Africa celebrated the Confederations Cup 2009™ and the first FIFA World Cup to be hosted on African soil. In the same year, YDF signed public–private partnerships with Volkswagen and Nike to increase YDF's impact. With the 2010 FIFA World Cup™ now on the doorstep of South Africa, 2010 was the busiest year for YDF in terms of special activities, events and visitors – the German Chancellor Angela Merkel, and UN Special Advisor on Sport for Development and Peace, Willi Lemke, being the most prominent of them.

More importantly though, 2010 was the year that saw the YDF Toolkit come to life when the YDF Manuals for Instructors and Coaches were published for the first time in June. From there on, the YDF Toolkit was developed further and short modules were developed and published in 2011 and 2012. While 2011 saw the implementation of YDF programmes across Africa in the spotlight, the YDF Manuals were granted patronage by the UN Educational, Scientific and Cultural Organisation (UNESCO), thereby acknowledging the exceptional methods developed by YDF and its partners. In June 2012, the South African and German governments agreed to continue on this joint journey and in June 2012 YDF was extended until December 2013. This extra time will allow the roots of

YDF to grow stronger and deeper into the communities and fortify the impact on young girls and boys participating in YDF programmes.

Looking to the road ahead, we are confident that YDF, in particular, and the whole sport-for-development fraternity can expect a bright future. Sport as a tool to promote development is being increasingly recognised and acknowledged by governments, relevant development organisations and the private sector alike. The role sport can play in offering youth brighter perspectives, in reducing violence, encouraging healthy lifestyles, fostering gender equality and promoting social change in developing and developed countries alike is immense. The importance of focusing on youth as a target sports group, but much more importantly as drivers of social change, cannot be emphasised strongly enough. They are the future in each and every society around the globe, and supporting their positive development will always put you on the winning team.

Stories from the Field just couldn't be published to a better point in time. The case studies collected by the University of Johannesburg offer a unique view on the positive change that sport can bring to the lives, not only of young girls and boys participating in YDF, but also to their peers, friends, families and communities. We hope that *Stories from the Field* will spread the belief that we have in sport as an innovative, effective and exciting tool to bring about social change; and we hope you enjoyed the narratives as much as we do in working to offer youth a better future.

Annexures

Annexure

Impact Summary

Cassie Ogunniyi (University of Johannesburg)

Monitoring and evaluating the GIZ/YDF programme has been a complex venture with the diversity of programmes spanning 10 African countries including seven South African provinces. There has been a multifaceted community involvement across a spectrum of government, civil society and corporate stakeholders. In addition to monitoring and evaluating programme impact, networking and network formation were also investigated. The complexity of the project is increased by indirect beneficiaries receiving Toolkit training, in addition to networking and the construction of networks such as Nike's Sport for Social Change Network to leave a sustainable structure and legacy on the African continent.

Through a process of scientific rigour, a monitoring, evaluation and impact assessment tool was developed, refined and validated to track the outcomes, intended and unintended consequences of the different partners as well as the YDF programmes as a whole. In the first instance, this system traces Outcome Monitoring Scores across 46 indicators relating to an overall objective, project purpose and four result areas with data collected through monthly, three-monthly and annual reports submitted by the partner organisations. This process dovetails with in-depth impact assessments that occur on an annual basis where the Most Significant Changes (MSC) are captured through site visits at community, institutional and individual levels.

As of 30 July 2012, 118,660 participants between the ages of seven and 25 have directly benefited from and participated in YDF programmes. Furthermore, at least 66,863 participants have benefited indirectly through YDF trainings and administrative support. This includes an additional multiplier effect of 1,565 Toolkit trainees from 375 organisations also implementing multiple programmes. This information is captured from 10 African countries, namely: South Africa, Botswana, Zambia, Lesotho, Namibia, Mozambique, Swaziland, Rwanda, Kenya and Ghana.

According to global trends of social impact assessments, the results were packaged to inform strategic decision-making (indicating challenges as risks to be managed and good practices that are duplicable). The responsible and responsive leadership provided by the GIZ/YDF head office led to refinement of indicators and optimal sustainability through capacity-building interventions in the field of sport for development as it intersects with health and educational development outcomes and practices.

Annexure II

YDF Impact Assessment & Annual Report 2012 – Executive Summary

The **aim** of the impact research is to provide contextual explanation of the recorded Indicator Monitoring Scores (IMS) and report on the Most Significant Changes (MSC) that are substantiated by a synthesis of questionnaire, focus group and interview data.

The impact assessment is focused on tracing the Most Significant Changes across 46 indicators in four **result areas**, namely:

i. Governmental entities and non-governmental organisations (NGOs) enabled to implement youth empowerment through sports initiatives,
ii. Youth empowerment through Sports for Development concepts, methods and tools developed and training of project partners is supported for implementation of the YDF-Toolkit modules,
iii. A system for the exchange of knowledge and experience has been set up and operates, and
iv. Sport-for-development events are used for educational, motivational and promotional purposes.

Currently GIZ/YDF have contractual agreements with 36 partners from South Africa (representative of all nine provinces) and in nine other African countries, namely in Ghana, Kenya, Rwanda, Zambia, Namibia, Lesotho, Botswana, Swaziland and Mozambique.

The revised Indicator Matrix (from January 2011) guided the dissemination of results and quantitative and qualitative data also is integrated with explanations of trends and specific dynamics according to the partner's programme mix, contextual realities and foci.

Regular monitoring and evaluation reports traced quantifiable data, and in-depth community level impact assessments took place in 2012 with 14 partners in five countries and four South African provinces, namely:

- South Africa
    - Eastern Cape (Imvomvo, United Through Sport (UTS) – previously known as Umzingisi)
    - Free State (MUCPP, SAFA Motheo)
    - Gauteng (Altus Sport, Dlala Ntombazana, SA Cares for Life)
    - Western Cape (Amandla Ku Lutsha, AYEP)

- Ghana (ACRO, YDS Ghana)
- Kenya (VAP)
- Namibia (YDF Namibia)
- Zambia (EduSport)

In 2011, ten partners in seven African countries were visited:

- South Africa (Soccer for Hope, Umzingisi, Altus Sport and Dona's Mates)
- Kenya (VAP)
- Rwanda (Esperance)
- Namibia (National Football Association)
- Botswana (SEDYEL)
- Zambia (Edusport)
- Lesotho (Kick4Life)

Qualitative information was also gathered from Yabonga (Western Cape) and MYSA (Kenya), as well as focus groups with members of YDF networks in Ghana, Rwanda and the Western Cape (South Africa).

A relatively complex research design allowed for pre-post comparison (from 2011 to 2012) for four partners, namely United Through Sport, Altus Sport, VAP and EduSport. According to the new direction of an intensified focus on training, "retrospective social impact" was also traced for network agency in Ghana and impact within ACRO (a network member). The research design thus allows for:

- pre-post comparison of Most Significant Changes over one year – between June 2011 – June 2012 (four partners),
- comparative cross-sectional analysis between partners, as well as determining trends within the funding cycle (13 partners visited, excluding Namibia),
- setting of new-baseline data across all stakeholders, and
- follow-up of case studies.

In 2012, a total of 993 research participants contributed to the different data sets that were integrated and discussed in this research report. The 177 interviewees included managers, implementers, participants and significant others of the latter two groups. Forty-four focus group sessions were conducted in which 258 discussants took part; they represented: network members (n = 6), implementers (n = 69), and participants (n = 113) with a gender composition of 57.3% male (n = 148) and 42.7% female (n = 110). A total of 653 respondents completed three types of questionnaires (differentiated between coach/administrator, secondary- and primary-school-aged participants): "Programme-focused" (n = 417) and "Lifestyle and behaviour" (n = 224), whilst 10 managers also completed Programme Information Data Sheets on behalf of their organisations.

The previous HIV and AIDS quiz was incorporated in the "Lifestyle and behaviour" questionnaire and administered to secondary-school-aged participants and also for the first time to implementers in 2012. Another deviation from the research instrument used in 2011 was to add more life skill-appropriate questions relating to the degree of resilience and to identify possible transfer for learning into everyday life and relationships.

The purposive sample selection also provided research participants (e.g. implementers, managers and participants) with an opportunity for data verification through triangulation of data sets. Follow-up interviews were conducted with 16 of the 45 comprehensive case studies collected during the 2011 research visits. Currently these stories are in the process of being published in which two versions are included – a more popular edited version of 15 stories in which a CD is included with a more academic version of the extended case studies. The publication, entitled *Stories from the Field: YDF Footprint in Africa*, is scheduled for publication in 2012.

Main Findings

The YDF project continues to progress according to the objectives set out at the onset of the programme and revised in January 2011, with most of the indicators increasing or remaining positive. The majority of the indicator values did not have a significant change compared to the 2011 report. A few indicators met and exceeded their target values over the past year, while a few decreased slightly and a few are less than 60% towards meeting the target. Indicators in these three areas will be discussed briefly.

There were six indicators that were met and then exceeded in the past year. The two most important indicators that were exceeded were the number of participants supported by the YDF project between the ages of seven and 25 in South Africa (**Purpose Indicator i**) and in the remaining project countries in Africa (**Purpose Indicator iii**). They both had a benchmark value of 50,000 participants. In South Africa currently 62,912 participants have been supported through YDF projects, while 55,748 participants have been supported in other African countries.

YDF started with a main YDF Coach Manual consisting of nine modules. Additional manuals were developed based on five of these modules and are considered short modules in: violence prevention, HIV/AIDS prevention, gender awareness, environmental awareness and sport event management. Training was a major focus for GIZ/YDF in the past year. This is reflected in three indicators related to training that have been exceeded. First of all, the number of coaches trained in the past year more than doubled (**Result 1, i**), currently with 1,565 coaches trained (IMS = 195.6%) from 375 organisations (IMS = 250%) (**Result 2, iii**). The coaches are from six South African provinces (Eastern Cape, Free State, Gauteng, North West, Northern Cape and Western Cape) and eight other African countries (Botswana, Ghana, Kenya, Lesotho, Mozambique, Rwanda, Swaziland, and Zambia). There have also been 183 Instructors trained from 71 different organisations from all the partner countries except Kenya (IMS = 116.4%). The aim for Instructors was 150 (**Result 1, ii**) which would result in an estimated 810,000 participants reached according to the cascade model. In the 2012 questionnaires and monitoring and evaluation visits, the quality of trainings was examined and it was found that 61.3% of implementers were satisfied with the level of training they received (an increase from 52% in 2011), which exceeded the indicator value of 60% (**Result 2, iv**).

Another area that increased significantly from 2011 was in the number of NGOs engaging with the YDF thematic areas. The benchmark for **Result 2, (ii)** was for 19 NGOs to engage with the YDF thematic areas, while there are currently 53 NGOs. Most of these newly added organisations have been across South Africa as GIZ has signed contracts with 16 new organisations since the past annual report and is now operational in all nine South African provinces. The majority of the psycho-social components continue to score beyond the benchmark of 50% of participants reporting improvements in the following areas:

- Enhanced self-esteem
- Improved acceptance of HIV/AIDS infected
- Transference of life skills to daily lives
- Increased pro-social behaviour
- Improved gender sensitivity and relations
- Improved knowledge and understanding of HIV/AIDS

The qualitative data collected through the interviews and focus groups continues to support the positive impact the YDF programme has on improving gender relations, HIV and AIDS knowledge and acceptance of people living with HIV and AIDS, although specific examples of the latter were sparse. Implementers were particularly forthcoming regarding their highly regarded status as role models and coaches in their communities. However, the lack of formal employment was a major detriment to their satisfaction of the programmes overall.

There are five indicator values that are less than 60% of the intended target amounts. The other three lower indicators relate to participants and locations. The IMS for **Purpose (ii)** increased slightly to 37.1% for indirect participants in South Africa (n = 14,833), revealing a challenge in collecting reports of participants that may be receiving YDF-linked material, but are not being included in the reports. This is especially true of the numerous club football coaches that have attended coach trainings, but are not part of the YDF partners and therefore not required to submit monthly and three monthly reports. Of the 1,000 youth desired to receive training in the YDF Toolkit (**Overall i**), there have only been 569 (IMS = 56.9%). The final low indicator is the number of projects and programmes operating in rural areas (**Purpose v**), with a benchmark of 50%; there are currently 26.8% in rural areas (IMS = 53.7%). However, this indicator has also increased since 2011 as a result of more projects in rural South African provinces (i.e. North West and Limpopo).

Good Practices

The main good practices are reflected in the areas of most significant change above and can be summarised in three main points:

- Nike and GIZ/YDF partnership providing opportunities for networking and exchange through the Sport for Social Change Network (SSCN) Southern Africa.
- YDF Toolkit training and resource development providing knowledge, capacity building and leadership opportunities, which infers status upon implementers.
- The focus on "youth" as drivers of change.

Recommendations

The recommendations that follow from this report focus on strategic sustainability considering the needs of participants, implementers and organisations as a whole. The main recommendations include:

- Address the socio-economic status of implementers by providing opportunities to further their skill sets and capabilities to improve their employability for a wider spectrum of employers.
- Ensure all stakeholders are involved and engaged in decision-making and feedback procedures.
- Address the lack of variety of resources through strategic partnerships, increased marketing, networking and fundraising.
- Enhance facility, equipment and events management practices.
- Implement user-friendly monitoring and evaluation.
- Foster and allow opportunities for practical and creative programme approaches for multi-partner benefits.
- Provide technical advice and leadership to support policy development and partner integration.
- Strengthen the operation of the SSCN with strategic directive and collaboration.
- Focus on issues of sustainability, youth development towards resilience, and socio-economic empowerment.

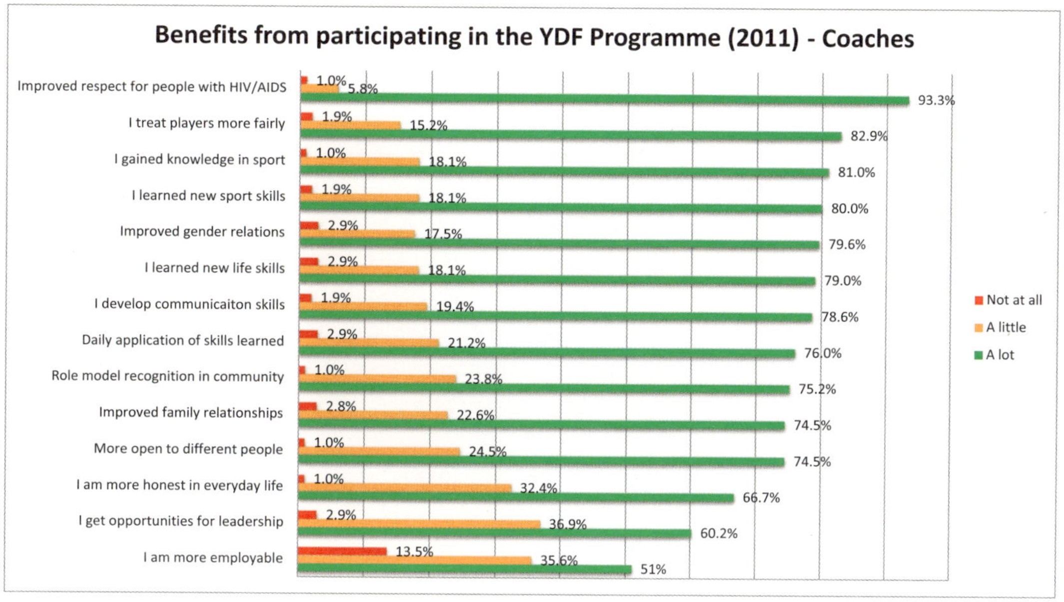

Benefits from participating in the YDF Programme (2011) - Coaches
Improved respect for people with HIV/AIDS
I treat players more fairly
I gained knowledge in sport
I learned new sport skills
Improved gender relations
I learned new life skills
I develop communicaiton skills
Daily application of skills learned
Role model recognition in community
Improved family relationships
More open to different people
I am more honest in everyday life
I get opportunities for leadership
I am more employable
1.0% 5.8% 93.3%
1.9% 15.2% 82.9%
1.0% 18.1% 81.0%
1.9% 18.1% 80.0%
2.9% 17.5% 79.6%
2.9% 18.1% 79.0%
1.9% 19.4% 78.6%
2.9% 21.2% 76.0%
1.0% 23.8% 75.2%
2.8% 22.6% 74.5%
1.0% 24.5% 74.5%
1.0% 32.4% 66.7%
2.9% 36.9% 60.2%
13.5% 35.6% 51%
Not at all
A little
A lot

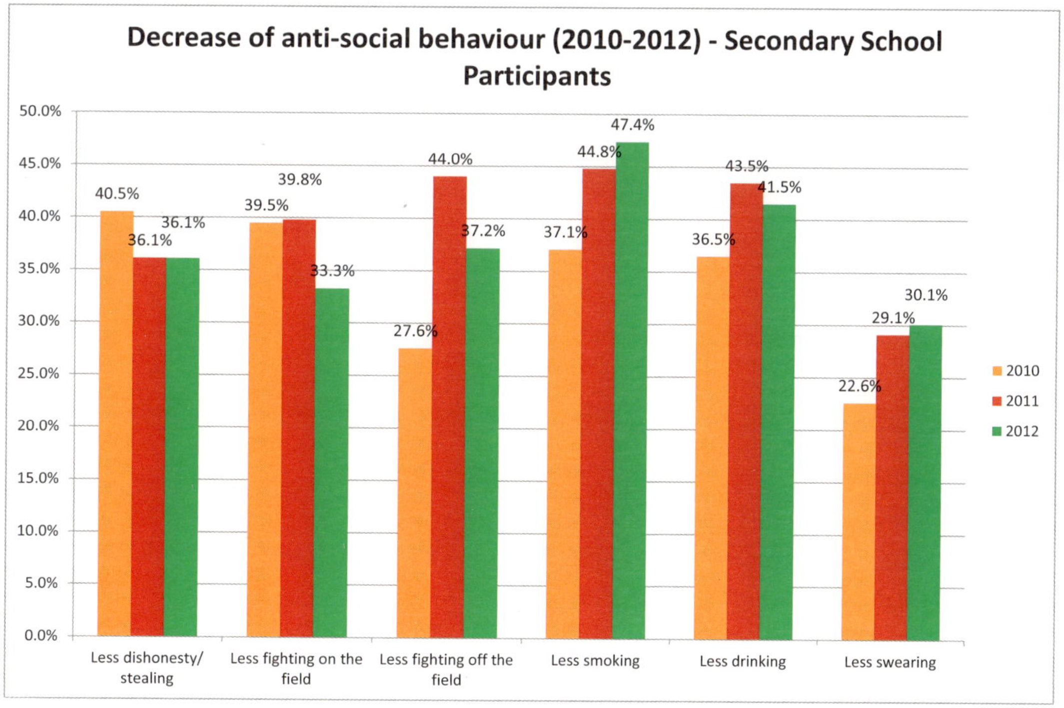

Decrease of anti-social behaviour (2010-2012) - Secondary School Participants
50.0%
45.0%
40.0%
35.0%
30.0%
25.0%
20.0%
15.0%
10.0%
5.0%
0.0%
40.5% 36.1% 36.1%
39.5% 39.8% 33.3%
27.6% 44.0% 37.2%
37.1% 44.8% 47.4%
36.5% 43.5% 41.5%
22.6% 29.1% 30.1%
Less dishonesty/ stealing
Less fighting on the field
Less fighting off the field
Less smoking
Less drinking
Less swearing
2010
2011
2012

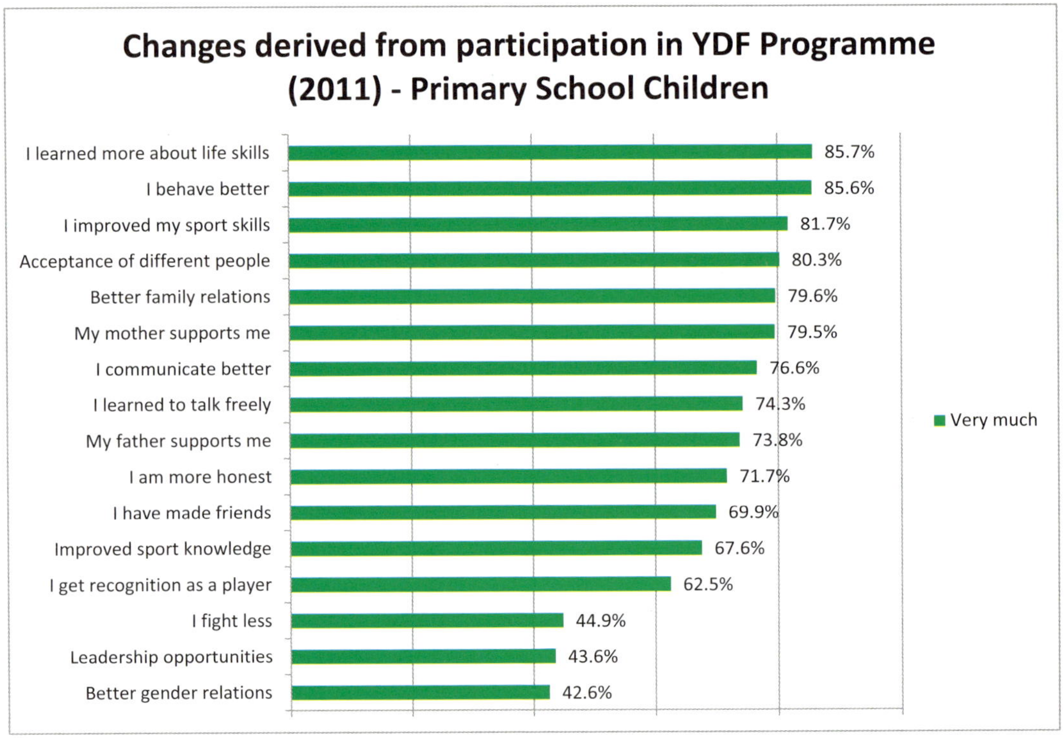

OVERVIEW OF INDICATORS

Overall Objective

Youth are included as drivers in social and economic development in South Africa and other African countries.

i. ≥ 1,000 youth are capacitated by the YDF Toolkit
≥ 1,000 youth are implementing YDF programmes

ii. ≥ 60% of youth report that they are recognised as catalysts for social change (e.g. as role models, provide leadership, change power relations, address local needs)

iii. ≥ 60% of youth implementers are exposed to opportunities (i.e. networks), experiences (sports-related) and training that will enhance their employability to be realised in the long term

iv. ≥ 60% of partner organisations acknowledge that youth contribute to the institutional capacity and sustainable programme delivery

Purpose Indicator

Youth are included as drivers in social and economic development in South Africa and other African countries.

i. ≥ 50,000 girls and boys between 7 and 25 years of age supported through YDF projects in South Africa
ii. ≥ 40,000 girls and boys between 7 and 25 years of age indirectly supported through YDF projects in South Africa
iii. ≥ 50,000 girls and boys between 7 and 25 years of age supported through YDF projects in other African countries
iv. ≥ 80,000 girls and boys between 7 and 25 years of age indirectly supported through YDF projects in other African countries
v. At least 50% of YDF programmes and projects implemented in rural areas
vi. ≥ 40% of participants in the programmes and projects are female
≥ 30% of implementers in the programmes and projects are female
vii. Government policies, plans and programmes reflecting YDF concept by mid-2012
viii. ≥ 3 countries with joint initiatives between government and NGOs targeting youth through YDF by end of 2011
ix. ≥ 9 countries with government involvement with NGOs targeting youth through YDF by end of 2011
x. 50% of the young people involved in, or reached by, YDF activities confirm that this has:

a. Enhanced their self-esteem and self-efficacy
b. Increased their pro-social behaviour in terms of demonstrating enhanced social skills, positive peer and family relationships with other household members
c. Decreased anti-social behaviour in terms of demonstrating less violent, unethical (lying, cheating) and discriminatory behaviour
d. Improved acceptance of people living with HIV/AIDS (reduction of stigma)
e. Improved knowledge and understanding of HIV/AIDS
f. Life skills are learnt and transferred to/implemented in every-day life experiences
g. Improved gender sensitivity and inter-gender relationships
h. Improved awareness and behaviour reflecting care for the environment

Result 1

Governmental entities and non-governmental organisations (NGOs) are enabled to implement youth empowerment through sports initiatives.

i. ≥ 100 MPP implementers trained in the YDF Toolkit by the end of 2011
≥ 800 coaches trained in the YDF Toolkit by the end of 2011
ii. ≥ 150 instructors trained in the YDF Toolkit (including short modules)
iii. 75% of partners express satisfaction with YDF measures
iv. ≥ 10 NGOs have measures and strategic plans in place for sustainable project implementation beyond 2011

Result 2

Youth empowerment through sport-for-development concepts, methods and tools are developed; training of project partners is supported.

i. ≥ 10 modules for the YDF-Toolkit are developed per thematic area, and ready for dissemination by the end of 2010. Modules comprise the combination of football with following youth-promotion areas:

a. Life skills
b. HIV/Aids
c. Health
d. Crime and violence prevention
e. Gender
f. Youth participation
g. Youth entrepreneurship
h. Youth with disabilities
i. Environmental education
j. Concept street football

ii. ≥ 19 NGOs engaged in at least one of the YDF thematic areas
iii. Trainees representing ≥ 150 organisations participate in training courses of Toolkit modules
iv. 60% of trainee confirmation of "good" or "accepted" quality of training
v. Instructors manuals exist for all Toolkit modules
vi. 25 partner organisations provide training on Toolkit modules
vii. YDF Toolkit is endorsed by relevant government authorities in South Africa

Result 3

A system for the exchange of knowledge and experience has been set up and operates.

i. ≥ 50 network members engage in at least one YDF linked network meeting per year
ii. SRSA or South African provincial government are actively involved in 75% YDF network meetings and exchange activities in South Africa.
iii. 75% of organizations benefiting from YDF network support are satisfied with the services provided
iv. ≥ 2 organisations per country have provided information on organisation's work (tools, lessons learnt, approaches, calendar of events etc.) to be accessible via the web-based information-sharing system
v. ≥ 60% of network members reported increased collaboration between members

Result 4

Sport-for-development events are used for educational, motivational and promotional purposes.

i) At least one public event per year which showcases the YDF concept used in each project country with a total participation of ≥ 100 active participants per event (2009 and after) and within South African provinces (minimum in 4 provinces), with a total participation of ≥ 100 active participants per event (2009 and after)

ii) ≥ 150 media exposure (e.g. paper, electronic, audio-visual) of YDF or YDF-related partners and institutions up to 2011

iii) ≥ 50 organisations (government agencies, NGOs, groups and individuals) receiving YDF support for events

iv) ≥ 6 countries mobilised to participate in events linked to YDF activities around World Cup 2010

v) By end of 2011, introduction manual for sport-for-development event management accessible for NGOs

Glossary of Terms

All Star: peer-leader, *see also* implementer, peer-educator.
AYDF: African Youth Development Fund.

BMZ: *Bundesministerium Für Wirtschaftliche Zusammenarbeit und Entwicklung* the German Federal Ministry of Economic Cooperation and Development.

Coloured: in the southern African context, a person of mixed race.
CWP: Community Works Programme, South Africa.

Dagga: marijuana.
Damara: ethnic Namibian tribal grouping.
DRC: Democratic Republic of the Congo.

EU: European Union.
EuropeAid: Directorate General Development and Cooperation–EuropeAid, a department of the European Commission (EU) operating under the authority of the European Commissioner for Development.
Evaluation: a process whereby the information obtained through monitoring is assessed to make judgments on outcomes or criteria and/or set descriptors that semantically differentiate between excellent, good, average/adequate or poor. It can also be expressed in terms of compliance with pre-set targets or indicators.
Event: a dedicated occasion (e.g. day of sport activities) used to spread the "Development through Football" philosophy. It is a once-off occasion occurring on one day or for a short duration. It may be repeated annually or more often but includes different participants each time. It does not include network meetings or training, but can be part of a larger project or programme.

FIFA: *Fédération Internationale de Football Association.*
FMF: *Federaçao Mocambicana de Futebol.*
Football for Hope centres: football facilities built by FIFA to promote public health, education and football in disadvantaged communities across Africa.

GIZ: *Deutsche Gesellschaft für Internationale Zusammenarbeit (GIZ) GmbH*, the German Agency for International Cooperation.

Herero: ethnic Namibian tribal grouping.
Hutu: the majority ethnic Rwandan tribal grouping.

Impact: or impact assessment, refers to measuring (quantitative data) or determining (qualitative) envisaged (intended and unintended) outcomes of a programme, project or event which can be expressed in efficiency (Programme Management and Delivery), social, human or individual and/or community development. Here the impact relates to emerging themes (qualitative) and/or can be expressed as MSC (Most Significant Changes) according to a pre-post design (comparison over time) and/or experimental control research design (comparisons between similar entities such as schools with, and schools without YDF programmes).

Implementer: a peer-leader, coach or volunteer who implements the YDF programme, *see also* peer-educator.

IMS: *see* Indicator Monitoring Score.

Indicator Monitoring Score (IMS): value of the current project indicator compared to the benchmark indicator values for a specific date and report; accumulative percentage of achieving a benchmark.

Indicator: a perceived outcome or pointer to show the quantity, quality or state of affairs (*see also* social capital, success indicator) of an activity, event or campaign. One could also differentiate between an output and input indicator where the latter refers to what is being planned or assumed as outcome without measuring the outcome or result related to the "input" (e.g. an activity, event or campaign).

Indirect Participation: the number of participants that are supported through an organization that takes part in YDF training or workshops, or is a regular member of a YDF network and is not a contracted partner. Participants also include those not directly supported by GIZ or included in the contract agreements, such as beneficiaries of programmes provided by people trained in the YDF Toolkit.

K4L: Kick4Life.

KESOFO: Kenya Community Sports Foundation.

Kgotla: a meeting of community elders.

LENEPHWA: Lesotho Network of People Living with HIV/AIDS.

Life Skills: social competencies.

Lobola: "bride price", a traditional southern African custom comparable to the European dowry custom.

M&E: monitoring and evaluation.

MDGs: Millennium Development Goals.

Monitoring: A process where information is collected regularly (e.g. three-monthly) and analysed to determine to what extent planned actions are followed in terms of the delivery of activities, aspects in place and/or changes required to achieve realistic outcomes. Indicator monitoring thus refers to the monitoring of activities and aspects with reference to a particular indicator as part of the planned Programme Management and Delivery component of projects and/or programmes.

MPP: Mass Participation Programme.

MSC: Most Significant Changes.

MTG: Moving the Goalposts Kilifi, Kenya.

MUCPP: Mangaung University of Free State Community Partnership Programme, South Africa.

MYSA: Mathare Youth Sport Association, Kenya.

Nama: ethnic Namibian tribal grouping.

Network: A constituted set of partners or partnerships that represents a particular brand or functional entity. It can be from an international body associated with FIFA (e.g. streetfootballworld or the Football for Hope Movement), a private–public partnership (e.g. Nike's Sport for Social Change Network). A network can thus represent a structure within a country focused to deliver YDF activities (e.g. Namibia YDF network and Ghana network) or within a region (Rwanda network in the Great Lakes Region) or province within a country (e.g. Western Cape YDF network). For the purpose of this book, there is a differentiation of an "in-country YDF network" and the Sport for Social Change Network (SCCN).

NFA: Namibian Football Association.

NGO: non-governmental organization.

Nyaope: a street drug, a cocktail of *dagga*, ecstasy and antiretroviral drugs.

Outcome: the final and eventual delivery that can be observed traced and assessed.

Participant: In this context refers to boys and girls between seven and 25 years of age who are the recipients and take part in the activities or events presented by YDF partners and or networks.

Partner: any non-governmental, governmental and/or private organisation or network that delivers YDF activities (e.g. programmes, projects, training, campaigns, etc.) in collaboration with and/or acknowledgement of YDF as an organisation and philosophy.

Peace Caravan: project supported by YDF in 2009 to promote peace-building and animate the idea of an African FIFA World Cup.

Peer-educator: an implementer, peer-leader, coach or volunteer who implements the YDF programme – an individual (mostly a youth) who has been contracted by an NGO to deliver a particular programme. Different programmes or NGOs will have categories within their programme that will differentiate between different types of "peer-educators" and might refer to them as peer-leaders, volunteers, technical assistants, All Stars or a term that will define their status and/or type of involvement.

Peer-leader: *see* peer-educator.

Programme: A structured developed set of activities or projects delivered according to the GIZ/YDF mandate; an umbrella term for the activities run on a regular, continuous and on-going basis; a programme can include multiple projects or events as well as other programmes.

Project: projects refer to different activities, events and/or community-based programmes implemented with the assistance of YDF. It is a time-bound set of activities with a specific purpose within a larger programme; each YDF partner might be considered as a project for GIZ/YDF central management.

RDP: Reconstruction and Development Programme, a South African government initiative to address the massive apartheid-era socio-economic imbalances, particularly housing.

Rural: having characteristics and components of a countryside, generally with low levels of governmental services, long distances to service points and low population densities; may differ from one context to another and will thus be identified in relativity to other settings by implementers within a particular region.

S•DIAT: Sport-in/for-Development Impact Assessment Tool.

SAFA: South African Football Association.

SASA: Scouts Association South Africa.

SEABE: Seabe Sports Development Association.

SEDYEL: South East District Youth Empowerment League, Botswana.

SEP: Society Empowerment Project, Kenya.

Shebeen: township tavern, often unlicensed.

Siyadlala: community-based mass-participation programme implemented in all nine provinces of South Africa under the auspices of SRSA (Sport and Recreation South Africa).

Social Capital: accumulated assets in the form of social networks, ties and/or liaisons that individuals, groups or institutions forge for gaining access to resources, or for the benefit of one or more partners.

Society: the community of mankind as a whole (global perspective) and/or an association of people united by a common aim, interests or principles within certain geographic or structural boundaries.

Sport-in-development: at one level it refers to the development of sport within a given context (Sport+ model) and the development through sport (+Sport model) where sport is a vehicle for community and/or human development.

SRSA: Department of Sport and Recreation South Africa, the Ministry for Sport and Recreation in South Africa.

SSCN: Sport for Social Change Network, a meeting point for organisations involved or interested in using sport as a tool for social development.

SSMPP: School Sport Mass Participation Programme, funded by SRSA and delivered by Provincial, Cluster, and School Sport coordinators at different levels.

SSTN: Swaziland Social Transformation Network.

Success Indicator: the recognizable result that presents the extent to which success or indicator benchmarks are achieved at the end of a programme, project or endeavour.

Tik: a street drug, methamphetamine; also known as crystal meth, meth, ice.

Toolkit: YDF has developed a toolkit around life skills and coaches' training which encompasses different levels (i.e. instructor and coach levels), as well as thematic short modules relating, for instance, to violence prevention and gender education.

TOT: Training of Trainers.

Township: in the South African context, a high-density, working-class settlement, implemented originally by the apartheid regime as dormitory towns for the black labour pool; townships are often poverty-stricken and lack basic services.

Tsotsi: a villain.

Tutsi: the minority ethnic Rwandan tribal grouping.

UJ: University of Johannesburg.

UN: United Nations.

UNESCO: United Nations Educational, Scientific and Cultural Organisation.

UNHCR: The Office of the United Nations High Commissioner for Refugees.

UNISA: University of South Africa.

UNOSDP: United Nations Office on Sport for Development and Peace.

Urban: having characteristics and components of a town or city generally with greater levels of governmental services, short distances to service points and relatively high population densities. It may differ from one context to another and will thus be identified in relativity to other settings by implementers within a particular region.

UTS: United Through Sport, South Africa.

VAP: *Vijana Amani Pamoja* (Youth Peace Togetherness), Kenya.

Volunteer: a person who serves voluntarily as a coach, administrator and/or assistant to a programme or service offered by community-based programmes and/or events. In some cases volunteers can receive some form of remuneration or stipend for subsistence and/or personal costs associated with programme implementation.

WCCF: West Coast Community Foundation, South Africa.

Xhosa: ethnic Nguni tribal grouping from the Eastern Cape province of South Africa.

YDF (*also* GIZ/YDF): Youth Development through Football (YDF) is a supranational programme implemented by the GIZ in partnership with SRSA (Sport and Recreation South Africa) and up to nine "other" African countries. The project is funded by the German Federal Ministry of Economic Cooperation and Development (BMZ) and co-funded by the EU (European Union).

YDF Coach: *see* peer-educator, implementer trained in the YDF Toolkit, to deliver life skills training through football.

YDF Instructor: *see* peer-educator, implementer trained in facilitating training for YDF Coaches

YDS: Youth Development through Sports, Ghana.

Zulu: isiZulu (an Nguni language) speakers with the highest concentration of population and traditional leadership in the KwaZulu-Natal province of South Africa.

Authors' Note

The case studies published in *Stories from the Field* are the result of academic impact assessment research by Prof. Cora Burnett, Prof. Wim Hollander and Cassie Ogunniyi University of Johannesburg.

For the collection of stories published in this book, a mixed-method approach was used within a Participatory Action Research framework (PAR) to ensure the local voices were heard and lived experiences captured. During the 2011 annual visits, a critical mass of data was gathered, comprising 931 respondents completing three types of questionnaires, 302 individuals participating in 31 focus groups, and 158 individual interviewees. The 158 interviews created 121 comprehensive case studies, comprising rich texts describing individual and community social world realities. The cases were narrowed down with 15 appearing here in print and 43 on CD, demonstrating the role of YDF as catalyst for change.

Stories from the Field is published in two versions. The present version represents only a limited selection of the case studies. The selected case studies were shortened and edited to match the specific audience and objectives of this publication. The unedited case studies are provided on the attached CD-ROM. All names of individuals mentioned in the case studies have been changed to ensure their privacy and personal security. Pictures do not in all instances represent individuals or locations of the stories.

All royalties derived from selling *Stories from the Field* are channelled to and shared equally among all the NGOs that participated in the research.

Photographic Credits

p6: UN photo by Mark Garten / p7: GIZ / p8: SRSA / p9: Thomas Ecke / p10: European Commission / p11: A. Proksch / p12: GIZ/Ralf Bäcker / p13: University of Johannesburg / p16 & 17: GIZ/Ralf Bäcker / p18: GIZ/YDF (L), GIZ/YDF (R) / p19: GIZ/YDF / p20: GIZ/YDF (L), GIZ/YDF (R) / p22: GIZ/YDF (T), GIZ/YDF (B) / p23: GIZ/YDF / p24: GIZ/Ralf Bäcker (L), GIZ/YDF, Simon Pause (R) / p25: GIZ/YDF, Simon Pause / p26: GIZ/YDF, Simon Pause (L) GIZ/YDF, Simon Pause (R) / p28: GIZ/YDF, Simon Pause (L), GIZ/YDF, Simon Pause (TR), GIZ/YDF, Simon Pause (BR) / p29: GIZ/Ralf Bäcker / p30: GIZ/Ralf Bäcker (L), University of Johannesburg (R) / p31: GIZ/Ralf Bäcker / p32: GIZ/YDF (L), GIZ/YDF (R) / p34: GIZ/YDF (L), GIZ/YDF (TR) GIZ/YDF (BR) / p35: GIZ/YDF / p36: GIZ/YDF (L), GIZ/YDF (R) / p37: GIZ/YDF / p38: GIZ/YDF (L), GIZ/YDF (R) / p40: GIZ/YDF (T), GIZ/YDF (B) / p41: GIZ/Ralf Bäcker / p42: GIZ/Ralf Bäcker (L), VAP/MYSA / p43: VAP/MYSA / p44: Moving the Goalposts Kilifi (L), Moving the Goalposts Kilifi (R) / p46: VAP/MYSA (TL), VAP/MYSA (TR), VAP/MYSA (BL), VAP/MYSA (BR) / p47: VAP/MYSA / p48: GIZ/YDF, Simon Pause (L), Kick4Life (R) / p49: Kick4Life / p50: Kick4Life (L), Kick4Life (R) / p52: GIZ/YDF, Simon Pause (TL), GIZ/YDF, Simon Pause (TR), GIZ/YDF, Simon Pause (BL), GIZ/YDF, Simon Pause (BR) / p53: GIZ/YDF / p54: GIZ/YDF (L), GIZ/YDF (R) / p55: GIZ/YDF / p56: GIZ/YDF (L), GIZ/YDF (R) / p58: GIZ/YDF (T) GIZ/YDF (BL) GIZ/YDF (BR) / p59: GIZ/Ralf Bäcker / p60: GIZ/YDF (L), GIZ/YDF (R) / p61: GIZ/YDF / p62: GIZ/YDF (L), GIZ/YDF (R) / p64: University of Johannesburg (TL), GIZ/YDF (R), GIZ/YDF (BL) / p65: GIZ/YDF / p66: GIZ/YDF (L), GIZ/YDF (R) / p67: GIZ/YDF / p68: GIZ/YDF (L), GIZ/YDF (R) / p70: GIZ/YDF (TL), GIZ/YDF (TR), GIZ/YDF (BL), GIZ/YDF (BR) / p71: GIZ/YDF / p72: GIZ/YDF (L), GIZ/YDF (R) / p73: GIZ/YDF / p74: GIZ/YDF (L), GIZ/YDF (R) / p76: GIZ/YDF (TL), GIZ/YDF (TR), GIZ/YDF (B) / p77: GIZ/YDF / p78: GIZ/YDF (L), GIZ/YDF (R) / p79: GIZ/YDF / p80: GIZ/YDF (L), GIZ/YDF (R) / p82: GIZ/YDF (TL), GIZ/YDF (TR), GIZ/YDF (B) / p83: GIZ/YDF / p84: GIZ/YDF, Simon Pause (L), GIZ/YDF (R) / p85: GIZ/YDF / p86: GIZ/YDF (L), GIZ/YDF (R) / p88: GIZ/YDF (T), GIZ/YDF (BL), GIZ/YDF (BR) / p89: GIZ/Ralf Bäcker / p90: GIZ/YDF, Simon Pause (L), GIZ/YDF, Simon Pause (R) / p91: GIZ/YDF, Simon Pause / p92: GIZ/YDF, Simon Pause (L), GIZ/YDF, Simon Pause (R) / p94: GIZ/YDF, Simon Pause (TL), GIZ/YDF, Simon Pause (R), GIZ/YDF, Simon Pause (BL) / p95: GIZ/Ralf Bäcker / p96: GIZ/YDF (L), GIZ/YDF (R) / p97: GIZ/YDF / p98: GIZ/YDF, Simon Pause (L), GIZ/YDF, Simon Pause (R) / p100: GIZ/YDF, Simon Pause (L), GIZ/YDF, Simon Pause (TR), GIZ/YDF, Simon Pause (BR) / p101: GIZ/Ralf Bäcker / p102: GIZ/YDF, Simon Pause (L), University of Johannesburg (R) / p103: GIZ/YDF / p104: GIZ/YDF (L), GIZ/YDF (R) / p106: GIZ/Ralf Bäcker (T), GIZ/YDF (B) / p107: GIZ/Ralf Bäcker / p108: GIZ/Ralf Bäcker / pp110 & 111: GIZ/YDF / p112: GIZ/Ralf Bäcker / p113: GIZ/YDF / p128: GIZ/YDF